A Flaw in the Ointment

A Flaw in the Ointment:
Explaining the Universe - The Difference Between How Things Are And How They Seem To Be

Michael Perilstein

Library of Congress Cataloging-in-Publication Data

Perilstein, Michael.
        A Flaw in the Ointment: Explaining the Universe - The Difference Between How Things Are And How They Seem To Be. - 1st ed.
        ISBN: 978-0-9712092-1-3

Published by Eala Dubh Editions for the International Society for Philosophical Enquiry, 700 Terrace Heights #60, Winona, MN 55987 USA

Printed by CreateSpace, an Amazon.com company, in the United States of America.

eala
Dubh
EDITIONS

Perilstein
Productions
TM

INTERNATIONAL SOCIETY FOR
PHILOSOPHICAL ENQUIRY
Quaere Verum

This book is dedicated to those who have understood and believed in me, especially Charlotte Perilstein and Penny August.

Contents

Prelude

I don't own a lot of things; so most people don't tend to pay very much attention to me.

Yet, I am a thinker.

When one spends most of one's time thinking, which is what I do best -- not that this is an indication of the *quality* of my thinking, but just that I can't do anything else quite as well -- it wouldn't be likely for that person to be a conformist (at least not mentally).

For the most part, critical and analytical thinking is not the pastime of the average person. Most people are average, by definition, and the average person tends to conform to whatever it is that the average person conforms to, and this does not include questioning the status quo very much, if at all.

The more rational and analytical one is, the less one tends to agree with many things that most people take for granted as being true, or else just don't really care about.

For example, I am a solid atheist (as opposed to being a liquid one?). I don't believe in anybody's god or gods, not even yours (if you have one). Also, I am totally convinced there is no possibility of life after death in the sense of one's conscious awareness continuing on to experience something that wasn't already experienced by that person during their lifetime. Neither do I believe, even for one moment, that the universe had a beginning, nor that it can have an end.

Oh, the Big Bang is true enough, but it's hardly the beginning. How do I know this?

That's one of the reasons why I wrote this book.

What else do I claim this book will prove?

Nothing can be infinite. (By that, I do not mean that there is something we call "nothing" and that this "nothing" is infinite. I mean, instead, that there can be no thing or things that are infinite—at least not in the real world, and at least not in the sense of being forever measurable without eventually returning to what has already been measured.)

I believe that I show, in this book, how it is the case that the universe itself has no beginning, nor can it have any ending nor can it have been caused. It has always existed. It is limited. It is finite. And the future and the past are already, always and forever, in existence.

There cannot be a 'god' that 'created' the anything at all, much less one that created the entire universe. Everything that exists must be physical. Probability is an illusion, as is possibility (that is, the only things which are truly possible are those which will happen or which are true). And we each experience our exact same lives forever (in exactly the same ways). That is the only "reincarnation" there can ever be.

Each of us lives our one and only very short life based on our views of reality, at least the causes of which mostly remain unconscious to many of us. Yet how do we formulate these views? Or, rather, where do these views come from? That is, how do they get into our heads and then become incorporated into part of our personalities?

In spite of what we'd like to believe, most people spend virtually no time at all forming 'their' views (or, rather, allowing their families, religion and society install these views into their heads). Yet they'll spend an entire lifetime defending them whenever 'their' views are brought into question.

Tell this to most people, and they'll argue endlessly – out of instinct. Shoot down 'their' views of religion, based on a series of logical contradictions for instance, and they'll come back with a totally different set of pre-conceived, pre-fabricated, logically contradictory "arguments." They'll be completely unaware that their reactions are just instinctive for the sole purpose of continuing to excuse/defend what they wish to be true, and what they intend to continue to believe anyway. This is especially odd since they may know, or at least suspect, that at some level, such ideas can't possibly be true.

In this book I present a long list of some ill-conceived ideas, which have

2

been taken for granted by many people over long periods of time, and then I proceed to tenderly, delicately, and deliberately demolish and obliterate them (the ideas, not the people), one at a time. Oh, don't be alarmed. I don't destroy these ideas because I'm mean; because, of course, I'm not mean. I destroy them because they're not true.

I have been arguing my ideas about these issues with very many individuals in person, on talk radio phone-in shows, and in newspaper editorials and letters (since the mid '70s) and then later on in "chat rooms" on the Internet (since the early '80s). As a result, the responses I received to my arguments were anticipated years before the individuals themselves were even presented with my arguments. I know how people think about these questions (not because I'm so great, nor because I have psychic powers, but because they've told me).

During such discussions, some people tend to feel threatened (even though they are not being threatened). And by instinct, and from reinforced past experiences, they react in order to defend what they want to be true. Yet it doesn't make what they wish to be true, true.

Be warned: in reading this book, your views of the universe and its contents may be changed forever. If these statements seem arrogant or self-serving, forgive me; I am just the messenger.

Judge the messages by their content and if you understand them to be rational and valid, then ask yourself how you can continue to believe otherwise, and why you'd even want to.

The subjects I deal with in this book, and the claims I make, backed by rational and logical argumentation, include the following (in no particular order):

1. We live in a universe in which everything happens the only way it possibly can, and therefore, there is really no such thing as what we call "free-will."

2. The "future" is completely determined. Not only that, but it happens exactly the only way it can which is already totally determined because it is *already* in existence.

3. The universe, and all of its contents, is permanent, It has it always been   here, was not "created," and can never be destroyed.

4. Your one life is permanent (as is everything else), and you re-experience it forever.

5. There are no "gods," nor "souls" outside of the concepts of such things, nor are there other types of actual reincarnation aside from the experience of one's own life perpetually.

6. Nothing can exist which is not physical.

7. Infinity does not describe reality accurately.   It is only a useful mathematical concept. Mathematics is a very useful tool but it has major flaws that make it forever impossible to accurately describe reality completely.

8. "Probability" is just an illusion.

9. It is not true that *anything is possible;* it's almost just exactly the opposite.

10. The meaning of life.

The purpose of this book is not to make you feel uncomfortable; rather, it is to get you to think logically.[1]

It's easy to be just like everyone else and follow the crowds. But it's pointless, for those crowds are just following the crowds themselves and, in fact, they *are* the crowds; so they tend to be following themselves into ever narrowing little spirals until, eventually, nothing is left (metaphorically). If you truly want to know about reality, then you have to start by examining

---

1. Naturally, this is not meant literally. A book, as we all know, is an inanimate object, and not a sentient being. Therefore, it has no interests, period. I, however, as its author, do. And it is my desire to inform the reader and make him or her (or it) think about things that are taken for granted but turn out, in fact, to be wrong.

everything you already *think* you know—the stuff you take for granted, which, perhaps, was never true in the first place. Then get ready to re-think it all over again.

As you read the upcoming essays on the various topics covered within this book, you may discover some things they have in common.

For instance, their content takes very little for granted.

They are highly critical. They penetrate into the heart of the problems they deal with. And, in doing so, prove that many of us tend to take far too many things for granted. We tend to blindly accept what turns out to be either completely false, only partially true, or, worst of all—totally meaningless.

Very few people tend to question whether their own beliefs are true, let alone whether those beliefs have any real meaning. Beyond that, almost everyone seems oblivious to the *likelihood* that perhaps all, maybe many, and at least some of their thoughts and most sacred beliefs could be completely devoid of significance. When we are comfortable with the answers, we rarely tend to look deeply into the questions themselves.

So if this book accomplishes nothing else, let it show the reader how to no longer take things for granted. Let the reader no longer believe in unsubstantiated assumptions—no matter how convenient or emotionally pacifying those thoughts may be.

Thinking as a hobby

Given human nature, it should come as no surprise that the idea of the closest thing to an all-knowing human being is the concept of the proverbial wise man sitting atop the Himalayas. In the minds of many, this guy has gray hair and is bearded (in that sense, like yours truly – to the degree that I have any hair left). Unfortunately (and I hope, *unlike* yours truly), he is of virtually no help to anyone— least of all to the poor slob who's traveled halfway around the world for some advice.

The wise man tells his eager student a tale about some apparently paradoxical situation wherein the inquisitor of the story is punished, or humiliated in some way for having asked the question to begin with. Why is he punished? He's punished because he wishes to use his intellect to understand reality, rather than to simply be. The moral of such stories is generally something both provincial and anti-intellectual; and as such, usually leaves the Zen listener in a transcendental state of confused befuddlement. (What is the sound of one brain not thinking clearly?)

Before you go accusing your author of being "anti-Zen," which is not completely true, consider this: Zen is not "that branch of philosophical thought dedicated to solving puzzles and figuring out the nature of things in the universe through the use of logic and reason." And that is why you don't use it for that purpose.

It does, in my view, have a useful purpose. But it isn't for solving problems and understanding the nature of the universe. Its most useful purpose is for therapeutic psychological reasons to allow the "Zenee" to relax, to appreciate the "now," whenever that Zenee feels like doing so, and to train one to get lost in "the moment" (whatever that means).

Unfortunately, charming as these philosophical puzzles may be, they tend to contribute to the popular misconception about rational thinkers. That is, those who value the intellect are somehow dim-witted. We see this with the frequent portrayals of scientists and other intellectuals who are often referred to as being "geeks" and "nerds." Although, on the one hand, they are acknowledged for being geniuses, on the other, they are frequently portrayed as being complete idiots. Of course, those who do not value logical thinking may not notice this major contradiction; yet the contradictions don't stop there.

That we should even have a term called "common sense," indicates the distrust of uncommon sense, as well as an occasional hatred the masses seem to have for the pastime of thinking in general, and intellectuals in particular.

After all, "common" sense is the opposite of the sense of the uncommon. That is, the person who stops to think things through carefully, *before* deciding what to believe, tends to be the exception not the rule.

Those who neither back up their arguments with logic, nor with evidence, nor with reason frequently use "common sense" as their excuse for not doing so. It is a term that appeals to the emotions of some people. It both contains the word "common," to which most people can relate, and implies that it already makes sense, therefore lacking the need for any further justification by simply declaring the sense into it, as if by magic!

If one can get away with claiming a statement is just "common sense," then why would one need to prove it to be true?

However, in my view: *Common sense is always common, but rarely sensible.*

Yet I must state that, in spite of appearances to the contrary, this viewpoint is *not* elitist. I am against any attitudes attached to any viewpoints that may tend to divert the reader from his or her job of analyzing the rationale of an argument based solely on its merits.

Whether one is an elitist or an anti-intellectual, if either, is strictly a matter of one's personal value system. Value judgments are emotional matters and, as such, fall into categories not directly connected to matters of fact—except that they sometimes interfere in the process of understanding.

For example, a person may have a vested emotional interest in believing in something. Often such a person will tend to ignore any facts or evidence to the contrary. This tendency is deeply ingrained and difficult to overcome. It's difficult for people to be consistently intellectually honest, independent minded, and willing to place a higher value on the *Truth* than on defending one's own viewpoint (which is a more emotionally comforting thing to do) when that viewpoint conflicts with the truth. Unfortunately, most people are, truthfully, not so noble.

If I were in front of you in person, you might ask me, "Why did you write this book?" I might answer you: Certainly not to con people out of their presumably hard-earned money. If I had wanted to do that, I would have written a fantasy book, filled it with the kind of nonsense that most people want to read (that is, what they already wish were true), and have given it a mystical name such as *The Oracle Of The Afterlife* (or some equally vacuous title). I would have re-assured the reader of the existence of a god, and of souls. I would have claimed that nobody ever *really* dies, but that those of us who follow the magic book as interpreted by the properly authorized authority of that book, somehow go to a magical heaven where all of their wishes and hopes and dreams come true. Perhaps I might have claimed that 'souls' keep being reincarnated into newer bodies to have better and more exciting lives, and that this life is just a 'test'.

I might have included a chapter or two extolling the virtues of so-called "out-of-body" (out-of-mind, to be more nearly accurate) experiences; or about that lovely bright light (an oncoming train?) at the end of the supernatural tunnel that appears in so many charming stories about "near-death" experiences. But I have more respect my readers than that. For when those topics appear in *this* book, they are analyzed thoroughly, analytically and *honestly*. This book supports no half-baked asinine ideas, no matter how popular they may be. (It's either completely baked or totally raw – nothing in between.)

Finally, even though these essays have individual topics and can be read independently, they are presented here in a specific order, for a good reason. Ideally, you'll read them in that order. Many of the ideas in the later chapters will be that much easier to follow. Occasionally, I'll explain the same things in several different ways. That's because different people process information differently. Occasionally, I will warn those readers who have already understood the ideas under consideration, to continue on to the next section without suffering the pain of unnecessary redundancy (or unneeded repetition). Also, many ideas are inter-connected. That is why I may only briefly explain something later in the book that had been written about in greater detail earlier on.

Thank you for opening your mind. Now let it be filled with truth, logic, reason, good will and perhaps a touch of some kind of tasty flambé.

Rethinking the Universe

"Just between you and me, things ain't what they seem to be."

Nothing

I thought I might start this book by writing about absolutely nothing. "What is your book about?" someone might ask.

"Well," I could answer, "it's really about nothing. Or at least a tiny part of it is."

But it isn't, because there is no such thing as nothing (apart from it being a word and a concept of course).

What would nothingness be? What *could* it be? Does it exist?

It doesn't, because it *isn't* (that's the best way to describe it). It can't exist (in reality, outside of the concept of it), because if it did, it would have to be *something*, which obviously would contradict the very definition of nothing.

"Nothing" certainly isn't anything, let alone *something*.

The very concept of "nothing" or "nothingness," owes its existence (as a concept, that is) to the fact that we know what *something* is. For without all of the *somethings* that there are, we would indeed truly have nothing. In other words, add up all the somethings, and you have everything. Take away everything, and you would have nothing (including yourself). Only that would not be possible.

So what does it mean to say something exists?

For a thing to exist, it must be made of something (energy-matter) and it must occupy time and space, and they must all exist together. In other words, time exists but needs space and energy-matter or it wouldn't, etc. To simply define "exist" as meaning, "to be," or "that which 'is,'" would be like defining a "New Zealand jazz apple" as being a "type of apple." If you haven't already defined what an apple is, that definition would be totally useless.

Similarly, claiming the word "is" to be a synonym for the phrase "to be" or "to exist," would likewise be evading the very nature of the question of what it means for something to exist, or to be, or to claim that it 'is'.

My definition is perfectly reasonable, and this is how I use it throughout this book. "To exist" means to occupy time and space and to be made of, or bounded by, something (energy-matter). Even the near-vacuum of any given area of outer space still occupies time and is bounded in space, needing things (made of energy-matter) with which to place those boundaries upon the location of this vacuum -- without which, the vacuum would not 'exist'.

Therefore, *to exist* is to occupy time and space and be made of, or bounded by, something which itself is made of energy-matter. *Otherwise*, it's not there, and it isn't even an 'it.'

It simply *is not*.

Nothingness, of course, is just the opposite, and does just the opposite. It neither occupies time and space, nor is it made of anything. Nor does it have any other characteristics aside from the imaginary quality of *not* being (which, in reality, only applies to faulty concepts outside of the concepts themselves).

Before you go claiming that this is "just a matter of semantics," let me stop you.

The fact that we must use language to communicate anything at all is itself a matter of semantics – but not in such a way as to invalidate the content of the communications. In other words, to the degree that semantics are involved in this communication, that fact is irrelevant to the content and substance of these arguments.

Nothing is less than being merely *no thing*. It is also not "being" of any kind.

In other words, it's not simply the lack of a "thing" – for example, one might consider the fact that as the word "noun" is defined as being "a person, place, thing or idea," we can see that by the standard definition of what a noun is, London, for instance, which is a place, is considered to be separate from being just a thing. Therefore, if the word "nothing" meant only the absence of a "thing," then the place called London, or at very least the concept of the place called London (as being a noun), would be able to qualify by definition as being "nothing," since it's separate from being a thing. Of course London isn't "nothing."

"Nothing" is less than simply "no thing." It is the complete absence of every, and all, things, ideas, places, persons, etc. In other words: if one can discuss "it" (whatever the "it" may be), then "it's" not nothing (though it may well be only a concept). Ironically, it is almost the case that merely having the word "nothing," and the concept behind this word "nothing," makes it more than what it is. Even the "concept of nothing" and the word "nothing" are far more than "nothing" itself could ever be.

This is clearly not to say that just because one can have an idea of something, such as a god for example, that this thing "exists" in any other way than as an idea (which, of course, is a physical thing located in the functioning brains of people having ideas, or in some other storage medium whether electronic or not, such as in books whose physical existence contains ink on paper wherein the ideas are encoded to be interpreted and translated by functioning physical brains at various points in time).

It is to say, simply, that there is no "nothing" for real.

On a related topic, Descartes' *"cogito ergo sum"* ("I think, therefore I am") is redundant. The *I,* who is doing the *thinking,* would have to have been there to begin with, in order to be doing the thinking. So his famous phrase would have been less pretentious, and more nearly accurate, had it stated simply, *"I am, and I am aware that I am, therefore I can think."*

Or, simpler yet, just: *"I am."*

For anyone claiming *to be* surely *must be* (or else such a claim would not have been able to be made).

Is Anything Possible?

Of all those popular, yet perhaps unwittingly arrogant and vacuous slogans, the ones that claim *"anything's possible"* and its various variations: *"everything's possible,"* and *"nothing's impossible"* seem to be among the most annoyingly devoid of content.

What is it *supposed* to mean?

Are we truly to believe that anything we can come up with is actually something that may really be true or really happen?

We take far too many things for granted – for instance, our vocabulary. We are often unaware of the way in which language forces us to think (and how some thoughts force us to use certain language in specific ways).

One of the consequences of really believing the phrase "everything is possible," in any of its incarnations, is the idea that if we can *conceive* of something, then it is actually *possible*; that is, it *could really happen*. But what would make this so?

The notion that, "if we can imagine something, it is therefore possible," doesn't follow logically. Simply coming up with an idea does not automatically guarantee its achievability. Likewise, thinking about something happening, neither automatically allows it to happen, nor guarantees that it ever *really* could.

[Also, if it were true to say that *everything's possible*, then there would be no such thing as the *impossible*; which, of course, would render the word "possible" meaningless, therefore making such a claim itself impossible. Many things that can be thought of we know are clearly *not* possible at all. For example, we know that, at least on the macro level, an object cannot both occupy a certain space during a specific period of time and not occupy that exact same space during that exact time – as they're mutually exclusive and therefore logically contradictory making them physically impossible! That is, it can't both be there and not be there for the same observer(s) at the same time(s). I'm assuming my readers are not so thick as to require additional examples of things that are clearly impossible. The list could be quite long, and the important thing is to understand that *not* everything is possible!]

So once we've acknowledged that some things are possible and others are impossible, the next question should be: "How does one distinguish between those things which are possible and those which are not?"

The word *possible* means that the thing to which we are referring *can* occur, or *can be* true. If it doesn't mean that, what else could it *possibly* mean?

If an exact assertion were absolutely *not* true, then it would be false to claim that it *can* or *could* be true. Similarly, if something absolutely *cannot* happen, then it is equally absurd to proclaim that it *can*.

Whether or not we know if it can or cannot happen is irrelevant to whether or not it actually can or cannot happen.

After all, events never happen in vague, nebulous ways, at least in the macro world. Things/events, always either really *do* occur (in some definite region of time and space) or they really *do not* occur in a given area of space during a specific period of time. That is, all *real* events always really do occur, within the universe, in *real, specific locations in time and space* – regardless of how we may arbitrarily label those areas of time and space (as long as we are consistent within our frames of reference and our measurements within those frames of reference).

Therefore, other kinds of "events" must only be imaginary. That's what distinguishes a real event from an unreal one. A real event is something that *really does happen*, somewhere in the universe, in a definite area of time and space.

Therefore, any conceivable event is really only truly possible (meaning that it only really *could* occur) *if, and only if,* it really does (eventually) occur. And everything that does eventually occur always does so specifically.

"Ah," you may ask, "but how do you *know* that any particular event will occur?"

To that, I would reply, "Who cares?" It doesn't matter, because that question is totally irrelevant. Whether one *knows* and is able to predict, with total accuracy, the outcome of any given event, has no bearing at all on the *fact* of that event's outcome.

For example:

Ignacio might have thought he had the "free-will" to choose from amongst the following events: going to the cinema, reading a book, watching television, or going to sleep, etc. Instead, he decided to visit his friend, Mark. That means that he did *not* go to the movies and did *not* read a book, watch television, go to sleep, or do *anything else* during that exact period of time in question. It doesn't matter what reason(s) *caused* him to see Mark. Since he *did* see Mark, then *that* was the event that actually occurred in that specific time and place.

So, how can one stick the event of "seeing a movie" in the exact time and space that is *occupied* by the event of "seeing Mark"? There is no room. Only one thing *can*, and only one thing *must* occupy the same space at the same time. [Although it would not be the *same space* if it were not at the *same time*.] If something doesn't happen at a certain place and time, then it can't have happened there and then, under those circumstances, and so was never really possible – no matter how it may have appeared.

Naturally, each event occurs within its own specific, exact and precise individual set of circumstances. Consider the possibility of your going to New York City – something you may think is something you may wish to do.

If *someday* you really do go to New York City, then the event of *"your going to New York City"* was/is a possibility – but only a possibility for the exact day/s and time/s when you really do go there. Similarly, if you never, ever, go to New York, then *your going to New York City* is always an impossibility – regardless of how it may *seem*, and regardless of when it is being conceived or talked about.

Even if the reason you didn't do something was because you "chose" not to do it, no matter why,then that "choice" became a motive which caused (forced) you to act exactly the way you did, which, in this case, was doing something *other* than going to New York.

Post Script:

I have had numerous arguments (usually friendly verbal ones) over several decades with very many different people on these and various other topics covered within this book. So I know from experience that some people will still not understand the full essence of my argument on this subject without some additional comments, while others of you already "get it," and will find any further arguments about this to be superfluous.

Therefore, if you have already clearly understood what I've had to say on this subject, feel free to proceed to the next essay.

This topic is closely related to the next one, *Free-will vs. Determinism*, as well as to many other ideas contained within upcoming essays. So it would be useful to understand it.

I hate to appeal to authority, and I will be very hesitant to do so generally. However, some people might not be willing to read, much less think about, what I have to say unless I can cite some authority figure who agrees with my thoughts on these concepts.

I will greatly limit resorting to such tactics, as I believe it is wrong to re-enforce this unjustified prejudice and dependency upon this form of intellectual laziness (not to mention that it's a logical fallacy – even though I just did). But I will make an exception here, and in the next essay, briefly, in the hopes that it may pry open some otherwise tightly sealed minds.

Albert Einstein knew that we live in a determined universe.[1] He had ongoing arguments with his fellow physicists, about the consequences of some of their interpretations of quantum mechanics. He would clearly have understood and agreed with my comments regarding *possibilities, probabilities,* and *determinism*, even though he did not phrase these things in exactly the

---

1. Einstein argued, to his dying day, that we live in a determined universe. That, in spite of the fact that his contemporaries totally disagreed with him, due to their strict following of the Copenhagen Interpretation of the Heisenberg Uncertainty Principle.

same way. If he had, there would have been no need for me to write these essays. But he did know we live in a totally determined universe.

In summary: it is not true that *anything is possible;* it's almost exactly the opposite. In reality, the only things that are possible are those that actually happen (or are statements which are actually true), and this is regardless of our ability to know in advance exactly what those things will be and/or when, and how, they will occur.

How is this the case?

Every event occurs in a given time and space (such as your life next Tuesday between three and five in the afternoon). Each one precludes any other event from happening there and then (that is, from occupying its specific time and space in the universe). For example, you may believe that it's *possible* for you to go to New York City next Tuesday. But either you *will* go there (in which case it *is* possible), or you will *not* go there (in which case it is *not* possible). There are no other apparent potential possibilities. That you do not know very much in advance what will happen has nothing to do with what will happen.

Space and time are occupied.

They are occupied by the exact events located within them. You can't have the Titanic sinking in different years throughout history, nor in different places. It sank when and where it sank – and that, like all other events everywhere, is forever occupying its own specific piece of time and space in the universe (from the same frame of reference).

Therefore, *something* specific and definite absolutely *will* occur in your life for example, next Tuesday afternoon between three and five—no matter how many *apparent other possibilities* you can imagine. [Of course, if you're dead before then, the events occurring will be in your death, not in your life, and will probably include your rotting underground a bit, or else just sitting relatively still inside an urn or casket somewhere.] The point is that only one *exact event* will *absolutely happen* there and then. And that is what is, and always was, *possible*.

Everything and anything else, no matter how easily thought of, no matter how it may appear, is logically (and therefore physically) *impossible!*

Just in case you have lingering doubts:

How can something really be *possible*, meaning that it *can happen*, if, in fact, it never does happen? If something *never* does happen, then obviously it can't happen, and in no way would it be true to claim that it can.

Why not?

Because, simply stated, things that don't happen, don't happen. It doesn't matter *why* they don't. The fact that they don't occur at any given time and space means that they do not fit into the category of things that "can" occur there and then, thus it's not true to say, ever, that it's possible that they *can* occur.

"Yeah, but you don't know that it's *not* going to happen," people often say. Again, my not knowing that something is not going to happen *does not* mean that it *might* happen. This argument is exactly like going into a room only filled with complete strangers. You are wearing a bag over your head (not necessarily because you're ready to have sex). You think that because the bag is over your head, and you can't see, that it's *possible* that there may be people whom you know in that room. But the room is full of strangers to you (at that point in your life). It is not now possible, and it never was possible, that these people were anything other than strangers to you (at least up to that point in time).

Ignorance does not make things possible.

The only reason you ever thought it was "possible" for them to be people you might know, was because of your lack of information.

This thought experiment proves that fact.

What you believe is, or is not, possible is directly linked to what you do, or do not, know with certainty – but that has nothing to do with what really *is* possible, as we've seen.

The room full of strangers makes it impossible for any of them to be previously known to you (by definition; you can't 'know' strangers). Yet that bag (the bag of ignorance) over your head makes you believe that it's truly *possible* that not all of these people are total strangers. And you would believe, based only on your lack of information, that such things were clearly possible.

But they wouldn't be.

Believing something is possible, only because one lacks information, is hardly a solid basis from which to form an opinion – much less to continue defending an especially stubborn, but false, one.

It may be true that something *seems like* it should be possible, but that's quite different from it actually *being* possible. Also, the very idea of something being *possible* presupposes the existence of the *impossible* by contrast. [Part of the problem in understanding concepts like these rests in the faulty notions we have regarding the universe.]

It seems like fun to believe that at any given "point" in time, that there are an "infinite" number of things which are "possible," and that we "choose" which direction we are going to take amongst the many "possible" different futures. However, every single word in quotes in the above sentence is a word with connotations that do not agree with reality.

There are no real "points" in time and/or space. Nothing is "infinite." And the only things that are truly possible are those that really do happen and those things that are really true.

Read on!

Free Will vs. Determinism

The legal systems of many countries, including The United States of America and The United Kingdom are based on the presumption that we all have "free-will."

The term "free-will," is meant to represent the notion that an individual's decisions are made solely by that person at the times when those decisions are made, and that not all of the components involved in producing the decisions were previously in existence (in one form or another) prior to the decisions themselves. In other words, free will cannot exist in a completely determined universe, as they are mutually exclusive concepts. In such a universe, everything in it, including all of our individual decisions, would have to evolve precisely the way it does, and all of our apparent choices would only be illusions, except for the ones that we actually did choose.

Why?

Because in a universe of strict causality, where actions are the results of causes which force those actions to happen, whatever you do throughout your entire life was always going to happen – and in exactly the way it does happen, where it happens, when it happens and how it happens.

Is that the kind of universe in which we live?

In fact, we do live in a universe of cause and effect. That is, with only two exceptions: light, and the universe itself (as a whole), as fully explained in *How Did It All Begin?* [1] Everything happens the only way it can. All events are determined. This applies to every single human action – from the germs spread through some uncouth pig coughing and sneezing into your face, to your not finding my humour to be funny, to each and every single human decision we make.

---

1. So as not to confuse you (and to be consistent at the same time), until you read about the two exceptions, just assume that everything is subject to the laws of causality. The exceptions will be explained in detail in a later essay herein.

There is no such thing as "free-will" of the kind where our thoughts, or actions, or decisions, are completely determined by ourselves alone in such a way that we truly could have 'chosen' differently.

Of course, we all do make decisions – though this is not what's in question. But when those are the only decisions we could have made, under each specific circumstance we were under when we made them, the notion of free will falls apart.

The fact is that we *do* decide things, and that we *do* make individual choices.

The question is whether our decisions and choices are always inevitable (thus no free will), or can things truly occur "randomly" such that we could have "chosen" differently under the same circumstances and thus truly do have free will?

By *free will,* we mean that an individual's decision is made by that person exclusively at the time that decision is made, and that not all of the components involved in producing that decision were previously in existence, in one form or another, forcing that decision to occur as it did occur before the decision itself was made.

If free will existed, it would not be affected by external factors. Therefore, it could not exist in a completely determined universe. That is because in a completely determined universe, everything in it, including all of our individual decisions and choices, would necessarily have to occur exactly as it did. If this were a determined universe, then all of our so-called "choices" would merely be apparent choices, all of which would be illusions, except for the ones that we really *did* choose. [Then again, how can we consider what was *not* chosen to have been a choice anyway? Isn't a choice defined as that which we *do* choose? Isn't "that which was chosen" the only real 'choice'? Everything not chosen would have been merely an apparent, or imaginary, choice anyway.]

In a determined universe, all the elements involved in each decision-making process would have to be already in existence, in one form or another (or at least they would have to be completely inevitable) in order to fit the definition of "completely determined universe."

They would have to be (from our point of view) continually evolving into that which causes us to make our specific decisions and/or take our specific actions.

There is a third school of thought, however, on this subject. It's what I like to call the *middle-of-the-road-agnostic* point of view. This is when a person cannot seem to get enough information to figure out a certain problem, so he arbitrarily refuses to decide and then claims that since he thinks he can't ever know, therefore it can't be known or that it's a little bit of each.

The middle road is "chosen." In this case, such a person might argue that, "of course many things in the universe, if not most things, are determined by the laws of nature. But (somehow, and always in ways never explained) not everything is determined – for example, human decisions." In this way, they will claim to have free will in what they realize is an otherwise completely determined universe.

They never show how the human decision-making process is (nor, even theoretically, how it could be) divorced from those very same laws of nature that control everything else. They simply make their proclamations and smile. No need for any evidence to back up their claims, because most of the time they are preaching to the silent choir anyway. They required no rational argument, proof, nor even any evidence whatsoever, to believe what it is they already believe, and certainly nothing at all in order to continue believing it.

Another argument in favor of free will is the religious one. It states simply that "god" gave us free will.

That's it.

Good, rational, thoughtful, intelligent argument, isn't it?

It assumes that there is a god. It assumes that this god is described, accurately, by *their* religion. It assumes that there is free will. Then it assumes that this god has the power to grant us free will – even though, as we shall soon see, such a thing would not be a logical possibility, and therefore would be physically impossible.

If this god were the "all-powerful, all-knowing, creator of everything," then it would certainly know in advance how all of its creatures were going to behave. (Otherwise it wouldn't be "all-knowing.") Yet if it knew how they would behave before they themselves knew, then obviously their actions would have been predetermined – in that case, by a god (after all, such a god could only know in advance what would happen if these things had been predetermined, otherwise no specific events would be happening in definite ways to know about). Hence, there would there no room for free will.

The religious argument is especially useless when one understands that there is no such thing as a god in real life anyway. Naturally, that doesn't stop many people from believing – most of them precisely because of a lack of proof of such a thing. They simply like the "god justifies the existence of free-will" claim. And that's the end of their argument.

[To these people this solution must be psychologically satisfying. They don't seem to care whether their views are logically sound, nor whether they are supported by evidence. Neither do they care whether such views make sense, even in theory. Nor do such people care to what degrees such claims do or do not correspond to reality. They seem mostly concerned with the apparent satisfaction they have achieved by finding a permanent place inside their heads for such ideas. And what's worse, is that they offer nothing at all in the way of support for them. That is, apart from their confusion over the issues – which, I suppose, they must assume applies to everyone else. They may even go so far as to act humble. It's as if there were a positive correlation between acting humble and being humble; and perhaps just as pitifully, between being humble and having the right answers.]

Perhaps one might speculate that the laws of nature have themselves given us free will. This would mean, absurdly, that they have *determined* that we have free will. Of course, that would be just another meaningless contradiction. For in a determined universe there can be no free will (by definition).

Yet, religious believers will simply believe in (or claim to believe in) free will anyway –regardless of the glaringly obvious irrational contradictions. Some will claim that their god says that there is free will, and so there is free will.

24

And that's that.

In such cases, we are simply told that we have free will, just as we are told that there is a god. We are given no evidence whatsoever (except for the evidence to the contrary which some very silly people will say was "put there by the devil," to fool us). Of course, they will claim the existence of a devil, too, with the same lack of reason, logic and evidence. And when one disagrees, these believers will insist that *we* prove to *them* that there is no such thing – as if, suddenly, and only for this one time, they value the notion of proof.

In other words, they can claim that something exists, or is true, and feel perfectly comfortable requiring no evidence whatsoever to support these claims. But when someone else comes along with contrary, rational opinions, those same people suddenly require something they had never needed before: evidence! And there can never be quite enough to cause them to change their minds.

As in the question of the existence of a god (see *The Folly of a God and the Problem With Religion*), we should first ask, "Who has the responsibility to prove any claim?" Or, who is obliged to prove a statement?

This famous age-old question has a reasonable solution.

The nature of proof is such that one cannot *prove* a negative (or so some say when what they had meant to say was that one cannot *disprove* a negative, both of which are false remarks anyway). One supposedly cannot prove that something does not exist because we cannot show you its lack of existence (see the section about "nothing").

However, one could claim that something exists in a specific location to which we could both travel -- such as the claim that there is some cheddar cheese in your kitchen refrigerator in a given period of time. Either cheese, or a lack of cheese, would be evident. In a sense, we could show it to be completely devoid of cheese. Still, true *cheese believers* could come up with another excuse such as, "It's there, but it's invisible."

Then we could point out the lack of cheese-like smells coming from the kitchen, to which one could say, "Well *I* can smell it, but *you* can't because you are not a true cheese believer," etc.   Such cheesy arguments are seemingly endless.

The point of all this is simply to say that plenty of things absolutely do *not* exist, and this can be proven to be the case, if even only indirectly, by showing that what is claimed to be in existence really isn't there (where and when it's supposed to be) and that something else occupies that exact space and time instead.

For another example, there are no live dinosaurs currently in my den. How can I prove that? After all, it's a negative. I cannot show you something that isn't there. But I could (in theory) take you into my den and show you what is there – and since no two things can occupy the same space at the same time, whatever is there precludes everything else (including dinosaurs) from being there during that time. You could always claim that they somehow really *are* in the room but that *I* can't see them. Or, that they exist, but not when anyone is looking for them (Or, if you really wanted to be creatively unreasonable, you could argue: "It's not dinosaurs that you've proven not to currently exist in your den, rather, it's Martians you have shown not to be there!") Or one might use countless numbers of other such arguments – of a more or less sublime nature. Most rational people would quickly see through such frail claims.

[There are many excuses available to those who wish their false beliefs to be taken seriously—no matter what the subject. This is true especially when there is no rational support whatsoever for such ideas. And sometimes these people may know, on some level, that their beliefs are ultimately groundless.]

The larger question is why should it be incumbent upon me, for instance, to prove that something that you claim exists, really does not? If you are the one to make the claim, you should be the one to back it up. The proof should rest on the side of those proposing the existence of a thing, or the validity of an idea. That does not mean that we can't argue effectively against false ones.

As stated before, one *can* disprove a negative.

If the idea is inherently flawed, irrational, and/or false, then we are able to "prove" it to be false. Of course, like beauty in the eye of the beholder, *proof* is always in the mind of the "provee."

Unfortunately, anyone can disregard legitimate evidence and the most reasonable and logically compelling interpretation of the facts. One can be perpetually stubborn and unreasonable (such as being a solipsist)[2]. However, repeated failed attempts to prove something to an idiot makes others wonder who the real idiot is.

In the case of free will, it is important not to make the mistake that some people tend to make. That is, we already understand that individuals do, in fact, make decisions. They *do* choose what they "want to choose" from all the apparently available choices.

However, they have no "choice" about what it is they "want" to choose.

What is it that *causes* them to want what they want, or to want to choose what they want to choose? The mechanisms involved in their *wanting to choose* exactly what they do choose, as well as the selection of apparently available choices, are ultimately out of the hands of the individuals making those choices. They are determined before any such apparent choices would ever seem to be an option.

We humans are part of the on-going process of *causes and effects,* and are not separate and apart from this process. So we act the only way we can, under every single circumstance, based on those laws of causality. The structure and function of our brains, and on our currently evolving environment, are all part of this causal chain of command. That means that everything that happens (every event) is caused to happen in precise ways by prior events having pushed that event into its exact location in time and space (from the point of view of

---

2. A solipsist is someone who'll waste his time arguing with someone he doesn't believe has an in-dependent existence. Also, a solipsist makes the claim that all of reality is subjective, since the only thing any of us can ever know, with absolute certainty, is that we exist (because we have perception). Therefore, if I were a solipsist, I'd foolishly try to convince you that, to the extent that you exist at all, it's only because I exist. The problem with this conclusion, dealt with in greater detail later in the book, should be painfully obvious now anyway.

the observer within the universe traveling, necessarily, at less than the speed of light).

I will demonstrate the irrationality of the idea of free will:

All things are exactly the way they are at any given moment because of the way they were the moment before. Similarly, things will be the way they will be a moment from "now" because of the way they are "now."

We understand this to be another way of saying what Einstein has said. That is, if one takes the concept of universal causality (cause-and-effect) seriously, then it follows logically and necessarily that this is a determined universe. It does not stop suddenly and wait for us to "freely" decide anything.

I would like to continue to show how there is no room for free will in a determined universe.

There are two factors that determine *everything* that happens to us in our lives, including *everything* that we do. One is our heredity (that is, everything from the outside of the surface of our skin, inwards). The other is our environment (everything from the outside of the surface of our skin, outwards). That covers it all.

If you currently believe that any *other* factors cause you to act exactly as you do, then think again. If you cannot clearly identify such factors, an intellectually honest person would abandon such an unreasonable belief. If you claim these other factors to be something vague and/or supernatural (such as "god" or a "soul"), this book should eventually help you dispose of the plague of those ideas once and for all (that is, if you value intellectual honesty).

Another problem arises when sloppy thinking fails to distinguish between *influence*, which we all have, and *control*, which none of us has.

Control means that we are solely responsible for determining the outcome of an event (such as one of our own decisions). Influence means that, although the event is causally determined, even before our birth, we are a (necessary) chain in the causal link. We are part of the mechanics of the actions leading to the outcome of any given event of which we are a part (including any decision of ours).

For example: a marionette puppet picks up a glass of water and drops it, causing it to break. We know that a human operates the marionette. Therefore we would say that although the puppet *caused* the glass to break, and had an *influence* on that event, it had no *control* over the event. Other factors *caused* the puppet to behave the way it did.

If we were to say that the puppet *controlled* the event, causing the glass to drop,' we would be wrong and arbitrary. Without looking any further, we would ignore what caused the marionette to act the way it did. And the same thing goes for the person behind the marionette and what caused him to act the way he did.

As part of the universe, we humans are subject to the same laws of nature as is everything else. If you doubt this, try "willing" yourself to be weightless, thus defying gravity.

Even if it *were* somehow possible for us to make a decision with no outside factors controlling it, we would have to acknowledge that our brain caused our decision (as that's where our thoughts are located).

Did you have anything at all to do with how your brain works, or what it's made of? Did you "choose" to be human? Did you have any influence upon your genetic make-up – before you were even in existence?

No, you didn't.

We had neither control, nor influence, over the structure and function of our brain, so it should be obvious that the output of our brain is similarly beyond our control. The strong illusion of control over the events of our lives may seem absolute. Naturally, thinking that we have more power than we do is comforting to many, though not intellectually honest and misleads us into the false belief that we have free will in our decision-making process.

What follows is an expanded example (taken from the previous chapter, *Is Anything Possible?*), of the kind of situation in which someone makes a decision, for reasons either known or unknown to oneself, causing the illusion of free will. We will see how such a person can easily fall into that trap.

Here's the situation: You have a free evening. You do not know what to do with your time. This fact, that *you* don't know in advance precisely how you *will* be spending your evening, is exactly what gives you the *illusion* of free will. It appears that you have the following choices:

1. You could visit your friend, Mark, at his house.
2. You could go see a film.
3. You could read a book (on philosophy, written by yours truly).
4. You could listen to music (no other comment).
5. You could watch television.
6. You could stare off into space while imagining yourself to be something other than just another peasant in an oligarchic society being exploited without your knowledge but with your unconscious and oddly enthusiastic consent.

Sound familiar?

You currently have no idea of what you *will* do. But, you will do something. That's certain. So you eventually "decide." You believe your decision was *"of your own free will."* You have decided to visit Mark.

Why? He has purchased some rare coffee from Indonesia for you and you wish to retrieve it. You miss talking with him; you have not had a good conversation with anyone for several weeks. He borrowed your dog and you want it back.

Pick a reason.

You can see the film you had been considering at another time, or you can read the book, listen to the music, or watch the same television show at some other time (pre-recording it, thanks to technology).

But you were more motivated to visit Mark.

And so you did.

Any motive, whether conscious or unconscious, *when acted upon,* is forever a part of the causal chain of events. [Of course, there are no such things as individual "causes" standing alone, as every cause is also the effect of a prior cause, but you get the picture.]

If, on the one hand, you are not consciously aware of your motives, you may believe that you haven't any. That just means that you need to get more in touch with yourself. If, on the other hand, you *are* aware of why you have done what you've done, you might boldly proclaim that your decision was "proof" of your "free will." You could just as easily have, for example, gone to the movies that night instead of visiting Mark.

But you couldn't have.

There were reasons that *caused* you to do exactly what you did, in exactly the way you did it. And if you could go back in time to the exact *same* situation, you would, necessarily, *be* in the exact same situation. Therefore, you would do the same exact thing in the same exact way (or it wouldn't be the same exact situation).

If you still find this hard to imagine, pretend we have a real life "videotape" that can be rewound and can replay any event. Unlike ordinary videotape, however, we would be rewinding reality. What would happen?

The same events would always unfold in the same ways for the same reasons every time. All things being equal, all things will be equal. Always. And *equal* does not mean *different*.

A note of caution:

Just because we do not have free will does not mean that we shouldn't act as if we did. Since we don't know in advance what *will* happen in many situations, we may as well act as though a variety of *apparent possibilities* are equally likely to happen. That's the way most people think anyway, although they may not know that there are no truly free choices.

How it is necessarily going to be the case that the illusion of free will must always appear to be real:

Even though everything happens in definite ways, the illusion of free will seems very convincing. This is not because things don't always occur exactly as they have to, according to the laws of cause and effect, but simply because we do not know in advance precisely how everything will occur. If we did know, this nice little comforting illusion would no longer exist.

I am going to show you, with a "thought-experiment," how it must always appear as though, even in a completely determined universe, there is free will. [A thought experiment is a device used by Albert Einstein for, among other things, demonstrating the validity of the Special and General *Theories of Relativity* without having to go into a laboratory where no experiments proving his theories would have been practical.] Yet, if we acknowledge a completely determined universe, such a thing as free will would have to be an illusion, even if a very stubborn and persistent one.

Let's *assume*, for the moment, something that actually is true: that the universe *is* totally and completely determined. In this universe we can also postulate the existence of a super-computer. Let's say this computer has access to every piece of information about everything, and is able to use this information to accurately predict the outcome of any event. So far this is logically consistent if not actually possible for technical reasons.

No offense, but suppose you were a trendy person. You find yourself sitting at a popular cafe with a friend, sipping some exotic blend of South Indian and Colombian coffee. This cafe has one of these marvelous new super-computers. So it has access to all of the information it needs to accurately predict anything. Still, not logically impossible.

Now suppose you were to ask the computer: "Tell me when monkey will drive on the door?" Of course, technically, you have formed a sentence, in the form of a question, but it's one that is completely meaningless.

The computer should reply with something like: "What the hell's wrong with you, are you insane?" Or, perhaps it might respond with the more computer-like comment, "Your question cannot be answered in its present form (because it makes no sense)" And, unless you are completely unable to function in the real world, you would understand this to be a very reasonable response.

Similarly, the computer knows that you are a stubborn and contrary person who will generally do the opposite of whatever you are told (unless you know that you are expected to do the opposite of whatever you are told – in this case, however, you don't know that). This is especially true when trying to prove the existence of your "free-will" (our computer knows everything, remember). You ask, "Will I finish drinking the coffee in this cup within the next ten minutes?"

Unlike the previous question about the monkey on the door, this one has true meaning. It does make sense, and it has a real answer. It is absolutely the case that only one of the following apparent potential possibilities will definitely occur: 1. You will drink the rest of your coffee within ten minutes. 2. You will not drink the rest of your coffee within ten minutes. (There are no other apparent possibilities – anything *else* that happens to the coffee will preclude it from having been consumed by you in those ten minutes. If you drink *some* of the coffee but not the rest of it, then you will not have drunk the entire cup.) Only one of those two (apparent) "options" is real. Only one can and must occur – the other one cannot and will not occur.

In this thought experiment, you *will* drink the rest of your coffee within ten minutes from the time of asking that question of the computer. That is what *will* happen – although it is entirely reasonable to assume that *you* do not know, at the time of asking the computer that question, whether you will or will not be finishing the coffee.

Now here comes the tricky part:

It's already a given that the events in this example are taking place in a completely determined universe—after all, that is the premise under which this little thought experiment takes place. You will finish drinking the coffee in that cup within the ten minutes of asking the computer that question. The computer knows that you *will* finish drinking the coffee. It also knows that you will do the opposite of whatever it tells you that you will do. That is, *if* the computer were to tell you what you *would* be doing, you would do just the opposite. So the computer *cannot honestly tell you what you will do,* even though *it knows.*

This is the only way to maintain logical consistency.

Therefore, although the computer can accurately predict the answer to this question, since it is "all-knowing," it cannot tell *you* the true answer. That is, it cannot tell you the true answer *in advance* (for that would create an "infinite (sic) loop" effect by making the computer's prediction become a part of the causal chain of events and creating a logical contradiction). This leaves the computer itself with only three apparent "choices":

1. Accurately predict this event, not for you, but for anyone else whose knowledge of the prediction would not be communicated to you in advance, and whose knowledge of the prediction would have no effect on the outcome of the event for whatever reason(s).

Or:

2. Not reveal this prediction. (Then, nobody would know that the computer knew the outcome, because the computer would have to keep predictions, of this nature, to itself. Admittedly, this is not very satisfying, but it is logically consistent.)

Or, lastly,

3. It could tell you that you will not drink any coffee within the next ten minutes. In other words, it could *lie* to you. It knows you will drink the coffee, and it knows that you would disobey and any predictions it made if you could. It also knows that its interaction with you in this manner would otherwise change the outcome of the event whose outcome is already known to the computer, but that this event's outcome cannot

be changed. So it could influence your behavior by giving you a false prediction. Again, this may be not very satisfying, but it is still logically consistent.

Some people might falsely conclude that the very fact that this computer has to take what you would do into consideration means that you have 'free will'-- which the computer already has to take into account. This is not so because all we need to do, in order to show a lack of free will, is to demonstrate that your actions are/were also caused. Furthermore, it follows necessarily that if what the computer tells you will *cause* you to behave in a contradictory way, then clearly you, too, have been caused to behave (or caused to "choose") in an exact, specific way.

In this case, the computer itself would have been a causal agent determining your decision making process and therefore interfering in your "free will."

[Don't pretend the computer itself has any "free will," either. Its actions, too, were totally determined by things outside of itself. And the computer has no say in the matter of how it operates – much like we humans.]

After all, there is nothing contradictory about having a contrary, obstinate person existing in a completely determined universe. In such a case, that individual's apparent choices, like everything else in such a universe, would still be *caused* by factors over which that person had no control – in this case, by the prediction of the all-knowing computer.

Remember, in this thought-experiment, we have already started with a completely determined universe as a given. We've added a character, you, who would do just the opposite of whatever our computer predicts (or rather, whatever our computer tells you it predicts). And our computer knows everything. Yet such a computer cannot, under every circumstance, always honestly tell everyone everything it knows.

When its very predictions enter the causal chain of events in a way that would influence the outcome of those events, such knowledge must, necessarily, remain restricted! Otherwise, we would have a logical contradiction (and therefore a physical impossibility).[3]

So, in the real universe, as in our logically consistent experimental thought one, the illusion of free will appears to be totally shatterproof.

Yet it is still only an illusion.

How to behave in a completely determined universe:

For many people, the fact that we have no free will is just annoying.

As mentioned before, the entire legal systems of Great (sic) Britain (it's okay, but I don't know what makes it "great") and The United States of America – as well as most other countries in the (so-called "free") world – are based upon the false assumption that we do have free will.

Also as previously pointed out, the fact remains that since we do not always know very far in advance, if at all, exactly how everything is definitely going to happen, we are left with the illusion of free-will, which is not always such a bad thing after all.

As long as we know it's just an illusion.

Since we will incur the consequences of our actions, we must take the moral responsibility for them (even though, in reality, we are no more responsible for how we act, technically, than a thunderstorm is for how it behaves).

Having no free will, in the true sense of the term, is no reason to excuse bad behavior. And, on the bright side, not knowing the outcome of events in advance gives us the ability to anticipate and to speculate and to gamble. It adds an air of mystery that, for some people, seems to make their lives a bit more amusing.

---

3. Some people would see this as a paradox (which, perhaps it is). [My definition: A paradox is something that contradicts itself, even though it doesn't.] However, I am thoroughly convinced that all paradoxes are only apparent. That is, all paradoxes can be solved through logic and reason. I will explore this issue in greater detail later on in this book.

The Illusion Of Probability

We have seen, in the chapter *Is Anything Possible?* how the idea of *possibility* is false when applied to the claim that something may be true of things that are *not* true and when applied to events occurring that never really do occur. The only things that are ever truly *possible* are those events that do occur and those statements that actually are true.

We have also seen how we live in a totally determined universe where absolutely *everything*, including all human decisions, happens the only way it possibly can. [An English language side note: Isn't it strange that even though the word "everything" is singular, its meaning is about as plural as it gets?]

Now we are ready for the next step: the illusion of probability.

In mathematics, the word "probability" means "the number of *possible* (sic) ways in which an event can (sic) occur, divided by the number of *possible* (sic) outcomes." This definition does not take into account the false assumptions that most people have already made regarding what is and what *is not* possible. Yet that's no small matter.

It follows that since nothing *can* occur which never *does* occur, then the idea that there can be a *probable* outcome, as measured mathematically, should be understood as merely being a useful tool for predicting the apparent likelihood of the occurrence of certain events based on what little information we have, and nothing else. Certainly it is *not* an accurate description of reality.

For example:

Event #1: take a standard coin such that it has an obverse side (heads) and a reverse side (tails) and "randomly" toss it. [By the word 'randomly,' it is meant that no conscious nor purposeful method for specifically and intentionally affecting the exact nature of the outcome of the event is intentionally underway.]

The standard prediction based on standard probability says that there is a 50% *chance* of it landing heads (and a 50% *chance* of it landing tails). We know that something definite will happen – even if it stands up on its side (which would make it a less than 50% *chance* of landing either heads or tails, but let's not worry about that for the moment).

The coin is tossed.

It lands heads.

The only reason we pretend that there is a 50% 'chance' of it landing heads, for instance, is because all we *know* in advance about this specific event, is that 50% of the coin is heads. [The other 50%, obviously, is tails.]

However, even in the language used, the word *chance* betrays the fact that there are *reasons* (that is to say, causes) why any, and in this case this particular, coin lands exactly as it does.

That we do not *know*, in advance, how an event of this nature *will* occur, has no effect on the specific and absolute occurrence of the outcome of that event.

The coin is unaware of our existence, and so it cannot possibly give a damn about what we think regarding how it will land.

But if we had all the information necessary to *know*, in advance, how that coin was going to land (for example: the weight of the coin, the angle, spin, momentum and pressure of the flip, air currents, etc.), then we would be able to predict, *with 100% accuracy,* the same event that others were predicting with only 50% accuracy. Yet they would be still be claiming that there was a "fifty percent chance" of the coin landing heads. We would have said, "No. The coin will land heads. It's one hundred percent probability."

And we would have been right.

Events in the macro world do not *partially* happen. No event only *half* happens.

This means that it's never the *event* that is uncertain (how could it be?), but always only the observer. All events occur in certain (definite) ways. The *only reason* we claim there is a *specific probability* that we attribute to any given event, is due to our lack of complete information regarding all related variables causing the specific outcome of that event.

Five people, with five different amounts of information about the same upcoming event, will make five different predictions regarding the so-called "probability" of its outcome.

Yet there will be only one outcome. There is nothing "probable" about it. The outcome will preclude any other outcome from having occurred and will prove itself to be the only real event happening in that exact time and space.

Event #2: A videotape of a horse race is shown to a previously sequestered audience. They are unaware of the outcome of that race. Yet they will claim the same apparent "odds" (or *chance*, or *probability*) as those who were there from the beginning, before the actual event took place. And those who know which horses will, *with certainty,* come in first, second, and third, are obviously able to predict this event with *100% probability.*

The apparent probabilities will vary, depending on the information known by the individuals at the time of making their predictions.

Thus, probability *is an illusion.*

It must be considered to be only an illusion. There is nothing magical about it either.

For example, in the World's Fair of 1965 in New York City, there was a demonstration of the principle of probability. There was a display, which consisted of two clear parallel plastic sheets separated by about an inch and a half. These sheets were imbedded into pieces of wood, along the bottom and sides, into which grooves were cut to fit the width of the plastic sheets. On the first piece of plastic, the one facing the visitors, was drawn a typical bell-shaped curve. This was the "probability curve."

The device would proceed to plummet a stream of marbles "randomly" until the bemusing bell-shaped curve, predicting where the group of marbles would land, had been filled-in – rather accurately conforming to the predicted behavioral pattern of these inanimate objects – thus proving the validity of the probability theory that the machine was designed to demonstrate.   T h e exhibit explained that through the miracle of mathematics, it could (accurately) predict where the aggregate of marbles would land, but that it was impossible to predict where each, or for that matter, any individual, marble would end up.

Post script: Probability is probability.

It's the same phenomenon whether predicting the weather, or a horse race, or coin flip (in the macro world)—or sub-atomic events in the quantum (micro) world.[4] It is exactly the same mathematical idea. This concept is inherently flawed when describing reality simply because it can only deal with large numbers of events but pretends to apply to specific individual events for which it assigns "probabilities" of specific occurrences (and always based on what little we know already). If correctly understood, it is only a useful tool to help us predict (that is, *to make better guesses* about) the outcome of events that always occur with certainty (that is, in certain exact ways).

---

4. Some of you are already getting annoyed with my apparent ignorance of quantum mechanics (the study of events in the sub-atomic world). Before getting too smug, think about the argument I just made and realize it applies to the concept of probability, no matter where or how that concept is used.

How Did It All Begin?

In The Beginning

In the beginning, there was no beginning.

Is the Big Bang real? Yes. Was it the beginning of the universe? No. Why not? Because the Big Bang is always only a *part* of the universe. You and I are also a *part* of the universe. But we are not part of the Big Bang (or we'd be quite dead). So both the *Big Bang* and *we* are part of a much larger universe. In other words, the universe was never just the Big Bang—after all, the universe contains all of time and space and all of its contents. [And how can something, which contains all possible real events in all different times and in all real locations, itself, have a "beginning"? What would that mean?]

I will emphasize this point later on as well, it *isn't* The Universe, if it doesn't include *all* of the contents of the universe – that is, including all of time, space and its contents (including the Big Bang and including us). We are not part of the Big Bang, but we are part of the universe. So the universe includes both the Big Bang and us. Therefore the Big Bang cannot be considered either as "all of the matter in the universe compacted together," as physicist Stephen Hawking has claimed (because *we* are part of that matter and *we* exist in the same universe but in a different region of time and space from where the Big Bang is located), nor as the beginning of the universe itself. I will show later on in this chapter, how the universe (in its entirety) cannot logically have been "caused," since the concept of causality cannot apply to it as a whole and still retain any meaning.[1]

If anything exists, ever existed, or ever will ever exist, it *is* (still) located somewhere within the universe that contains it.

---

1. I know that this claim seems quite odd, and as yet unjustified. Please continue reading and this will be explained later in the chapter – to do so now would take us on a needless tangent which would serve more to confuse than to enlighten at this point in our discussion.

So why do some otherwise well educated and reasonably intelligent people, many of whom call themselves scientists, try to figure out how the universe "began," without first questioning whether it could have even *had* a "beginning?"

They start with an assumption: that everything *has* a beginning because, somehow, everything has to have a beginning. This idea comes from the notion that in the causal chain of events if we go back far enough, there must be a *first cause,* or an original, uncaused cause.

Why there must be a first cause is unclear.

Perhaps it's because it would seem, on the surface, that if there *weren't* such a thing, then we would be backtracking causes *ad infinitum*. We would never get to the heart of the matter if we were unraveling a never-ending metaphoric ball of causal string. [But even if that were the case, it would only make the idea of not having a "first cause" inconvenient and not automatically untrue.]

Thus came the excuse of a "god" as the first cause, the un-caused cause, or as the "creator" of the universe. It is, supposedly, a universe that had to have been created, according to this way of thinking, because everything has to have a beginning (other than this "god" evidently).

Of course, this contains the very obvious major contradiction that cannot be ignored.

The very justification for the existence of such a "first cause" (or "god" as "creator" of the universe) is based on the *necessity* that *everything* has to have a beginning. Therefore, a god caused the universe to begin. The problem is obvious: this god would also have to have been created (or, have had a beginning) and for exactly the same reason(s).

Now we are back to where we started from, not that this is a new idea, nor one that hasn't been explored by many theologians and philosophers in the past.

On the other hand, if, somehow, this god didn't need to have an absolute beginning—if this god could have always existed—then one can no longer claim that, "*everything* needs a beginning." And the same reason(s) that could justify an ever-existent god, could more easily serve to justify an ever-existent universe. And that gives us a universe with no (absolute) beginning, one that has always been here.

Therefore, there is no room for a "creator" of something that has always been in existence; in short, no room for a god!

Yet it is not true that everything has to have a beginning. This is faulty reasoning. Again, what most people take for granted as being obviously true, is false. Nothing has an absolute beginning—as we shall soon see.

It *is* true that everything has to have a cause (provided that it is part of the universe and travels at less than the speed of light – more on that later). And that's where the confusion comes from.

On one level or another, most of us know that we live in a universe of cause and effect. However, lazy thinking has *caused* us to believe that a cause is the same thing as the *beginning* of whatever it is that follows that cause.

But it is not.

It is a *continuation* of the process of all that has gone before and all that comes after (universal evolution, if you will).

Unfortunately, even some scientists (that is, people who are otherwise scientific in their thinking) have made the same mistake when asking: "How did the universe begin?" They answer that—not with a whimper, but with a *Big Bang*.

The *Big Bang* theory, which comes from implications inherent in Einstein's *General Theory of Relativity,* is said to be "the beginning of the universe," in which state the universe was compact and condensed as much as possible (much like the universal contents of an astronomical trash-masher), and from which a giant explosion evolved into the stars and planets and, eventually, ourselves and even this book (and all that comes after).

It's cute, and there seems to be plenty of evidence to support such an occurrence, in the form of what is known as the *Doppler effect* or *Doppler shift*. [In light, a star moving away from us would be shifted (from white) towards the red end of the color spectrum (called a "red shift"). Stars moving towards us would be shifted towards the blue end of the color spectrum (called a "blue shift")].

Since we see "red-shifted" stars, in all directions, we know that they are moving away from us. Therefore, they were closer to us in the past. If we go back far enough into the past, the theory goes, everything in the universe would have been so close together that it would have, at one time, been so condensed, so hot and so compact that eventually it would have exploded—producing the stars and planets, etc.

One problem with this theory is that it postulates the existence of what is called a "singularity;" supposedly, the "entire universe" condensed into one spatial "point," which has no dimensions, and at which point (no pun intended), the known laws of physics apparently break down, making it impossible to predict what happened *before* the explosion. But as we'll see in the essay, *The Problem With Infinity,* points have no dimension and therefore cannot exist anyway (no matter how mathematically useful they may be on occasion).

It is also frequently said that we can't talk about a time "before" the *Big Bang* [the assumption being that the laws of physics break down at that 'point' (the 'singularity')].

But that's not true unless we assume the *Big Bang* itself to be an uncaused cause. And that is neither rational, nor logical, nor is it supported by a careful analysis of the evidence.

Here's the problem: as stated earlier, the Big Bang is only a *part* of the universe. After all, the universe is all of time and space and its contents (that is, energy-matter, which, according to physics, is spatially extended. That is, space cannot exist independently of the things occupying that space.). We cannot separate anything that happens within the universe from the universe as a whole. That is, we cannot remove any of the contents of the universe. The universe is *all* of reality. It's *everything* (which is real). If anything exists (or ever has existed or ever will exist), it is in, and forever a part of, the universe.

We know that time and space are inseparable—you can't have one without the other. [This is an established fact, but for those who are unaware of this, I will explain. Time is a dimension. It is the measurement of motion (internal, external, whatever kind—it doesn't matter). In order for something to move through space (or even stay in "the same" space *over a period of time*), time is required. Similarly, the passage of time requires some kind of activity or event to be occurring in space. You can't have one without the other.] So physics, under Albert Einstein, has combined the two into what it now calls *"space-time."*

The entirety of the universe is: all of time, as well as all of space, and its contents, (which can only be made of energy and matter, or what physics calls "energy-matter," and the nearly empty vacuum of space).[2]

Of course, I know that one can't simply define "the beginning" of the universe out of existence and then use that as evidence to prove that it had no beginning. But *if* the universe *had* a beginning, then it must follow logically that at one point (that is, *before* it "began"), it wasn't here at all. And that means it would have had to pop into existence from *nowhere*. But as we know, there is no such place as "nowhere." It doesn't exist. It is the same thing as "nothing," which also *isn't*.

How can *something* have come from *nothing* when, for among other reasons, there is no nothing for it to have come from? [Note Well: Don't let people throw you off course during a philosophical discussion by telling you *"You are arguing semantics,"* as previously pointed out (ironically, a grammatically incorrect, as well as a logically incorrect response). The use of all words is, necessarily, a matter of semantics, and that is why they should be used with precision and clarity of thought and purpose. It only fogs the issue when using words but not dealing with specific details. So, don't allow yourself, or anyone else, to reject some concept out-of-hand as being merely "a matter of semantics." That would be running away from the thought process in cowardice— denying

---

2. The vacuum of space is not "nothing," and it's not truly empty, for it contains field. That is, it can allow the transmission of electromagnetic radiation (light). And even the nearly empty vacuum of space is still unable to exist outside of time and space itself, and also requires its outer boundaries to be made of energy-matter, within which the vacuum itself is contained.

yourself, and others, the pleasure of understanding things you might otherwise never get to learn.]

Perhaps the universe didn't come from "nothing" and thereby violate logic and clear meaning by suddenly appearing. Perhaps it existed, in some way, in some other form.

But if the universe existed in some other form, then we should be asking the question "How has the universe *evolved?*" which is quite a different question from "How did it begin?" In fact, this is the question that true scientists should have been asking themselves all along.

At the risk of repeating myself, but for the purpose of making my point unambiguously clear, I will, in fact, repeat myself:

The universe as a whole is equal to the sum of its parts. The contents of the universe are made of energy-matter occupying space-time. And neither energy nor matter can be created nor destroyed (this is a physical law of Thermodynamics). Add up all of the contents of the universe, and you have the universe as a whole. Therefore, the universe as a whole could not have been created and cannot ever be destroyed.

Imagine the destruction of the universe. To *where* could we send it (or any part thereof)? If it were to be sent to "another universe" (a meaningless term), or anywhere else for that matter, then it would hardly have been destroyed. It would just have been relocated.

How are we to imagine the absolute destruction of *anything*?

Things change (universal evolution). Matter changes into energy, and vice-versa, on a sub-universal scale. But nothing ever goes away. Just as nothing ever simply "pops into existence," neither can anything simply disappear. Even in the misunderstood world of quantum mechanics (dealing with things on the sub-atomic scale), the physical law of conservation of mass-energy is still always, ultimately, validated (even if it occasionally takes a brief apparent detour into the *Twilight Zone*).

Also, if there had been such a thing as a "time before the universe existed" (again, a logical contradiction)[3], then we must ask ourselves another very serious and more perplexing question.

If the universe were ever not in existence (a truly empty concept, since time and space *does* exist), what would have been in its place?

The answer is: Nothing.

There would have been no place for anything to be. "Places" are located within the spatial and time dimensions of the universe—that's part of what the universe is: places (space) and time(s) along with energy and matter.

Furthermore, it is absolutely meaningless to say that anything can exist somewhere other than in the universe (remember, I've already defined what I mean when using the word "universe," as being all of time and space and all of its contents).

There is no existence apart from a physical one within space and time.

How could there be? What would that mean? In order for a thing—anything—to exist, it must be located somewhere (in space) during some period of time, otherwise it isn't. How could it be? What *else* would it *conceivably* mean to claim that something exists?

All that exists is always physical.[4] Human emotions exist in human brains, when they are being emotional. That is, the emotions occupy the time, during which someone experiences them, and the space, located inside the functioning brain, where the emotions are being experienced. And the brain activities, resulting in the feeling of emotions, are carried out by components of the brain that are composed of matter and energy, which makes up the brain

---

3. It's a logical contradiction, because it refers to time existing before there was a universe in which it could exist. It would also imply that at one "time," the universe would not have been "here." But time is part of the universe, and can't exist separately from it, as is equally true of location.

4. That is, it has measurable properties of matter and energy. Even "mental" activities are physical actions occurring in a physical brain.

itself. [For example, a brain is made of carbon and hydrogen and oxygen, etc.]

I know that this isn't the most well received concept. Many people will postulate "things" which have no substance (that is they are immaterial, or not made of anything physical) "existing" without being located anywhere. However, that is strictly meaningless. They want there to be some*thing* that is real and yet every single quality that defines real things is totally out of reach in this phantom world. These imaginary non-entities which can't be located, can't be shown to be made of anything physical, and in fact cannot even be detected indirectly, have absolutely no characteristic of anything real whatsoever. However, they do fit very well into the definition of the word "nothing" and their location fits well into the definition of "nowhere" – both of which we have already clearly understood do not and cannot exist. Of course, the odd thing is that if these imaginary non-physical 'things' had any real characteristics of the physical world, then they would be, by definition, physical.

Ah, irony! Isn't it wonderful?

At this point in the conversation (when I am boring someone in person with these same topics), I find that many people can't help themselves from re-asking the question, "Yeah, but what was there *before* the universe began?" Evidently these people are slow (at least to change their views). After all, especially as it is the case that the universe did not come from "nothing," they ask, where *did* it come from (as if it "came from" someplace "else" and wasn't always where it always is)?

Again, the "Big Bang," which is a real event, is generally labeled as being "the beginning" of the universe. The scientific community accepts the reality of the Big Bang. (The religious community may not agree, but it's out of their area of expertise, if they have any, as it is clearly a scientific issue requiring logic, reason, evidence and a degree of intelligence to decipher).

When the material in that region of the universe was as hot, dense and compact as anything can ever be, it all "exploded" (or, more nearly accurately, imploded in on itself due to condensed and compacted gravitational forces). Then it expanded, through the region of time and space that it caused by its very expansion, evolving into the galaxies, stars and planets, etc. And, on this planet at least, it caused the evolution of pigs and people and Cuban cigars and potato chips and motion pictures, and toilets, and religious fanatics, amongst other things.

I've explained, at least in some ways, how the universe has to have always been here.

Clearly, it never "began." How could the Big Bang be the beginning of something that can't have had a beginning?

Think of it this way: How old are you?

The answer to this question depends upon *when* you are reading that question (or when someone asks it of you). In other words, how old you *"are"* depends on *where your current conscious awareness is located in the universe* when that question is being asked of you. Let's say that right now you are twenty-one years old (a favorite age of mine). Let's assume you'll live to be at least seventy. Oh, hell, this is a thought experiment. Let's be nice and give you a life span of 120 healthy years.

You were once ten years old. When you were ten, you were still located (space-wise and time-wise) somewhere in the universe—the same universe we're all in "now." No force on Earth, nor anywhere else, could have ever removed you (nor anyone, nor anything else) from this universe—no matter how amusing the thought may be.

What are you aware of *now?* . . .

How about *now?* . . .

And, *now?*

Your current *conscious awareness* changes through time, but that ten-year-old you is still where it always was. You, at age ten, are located in a different space and time from where you are now. But you are 'both' in the same universe. That ten-year-old you is still where it always was.

Conversely, somewhere else in the (same) universe, you are fifty, sixty, ninety, etc.

Somewhere else, it is a million years after you died. All of these parts of space-time exist in this (the only) universe. Just as is true for the age question, what is "*now*" depends upon where you are in the universe when you consider this question.

Nothing removed you, nor could ever remove you, as a ten-year-old, or as an infant, or as the old person you have yet to become, from the very universe of which you are an incredibly infinitesimal, but absolutely permanent and integral part.

The infant you once were, you still are, where you were always an infant, in the time and space your infancy occupies in the universe. That infant is always where it is. Where else would it be? Where else could it be? How could it have gotten there?

We move away, time-wise, from what we call the past and into what we call the future. Where they both meet, we pretend is the present. It's what we call "now," although each time we define what is happening in what we call "now," we are instantly forced to re-evaluate our conception of what now "*is*." This is not simply amusing. It is a fact with both profound and hidden consequences, which cause us to view our one brief encounter with a fraction of the universe (that is what we call our life), through a pinhole. It is this very narrow frame of reference, in which we find ourselves, that confuses us when it comes to seeing the big picture of exactly what the universe is. After all, the point of view of a pinhole is hardly expansive.

For example, we know that everything that happens can be considered as an individual event. Any individual event is, of course, an arbitrarily separate occurrence within all of time and space (that is, within the universe). This is true because all events are located in time and space, and, when added altogether, equal the universe.

The freedom we have to move about space-wise is not extended to us time-wise. We move in one direction time-wise—forward (into what we call the "future")—and cannot go backwards.

Imagine if it were reversed:

You could walk into a room, but could only move forward. Everything behind you becomes totally inaccessible. Oh, as a consolation prize, we'll let you move about freely in time, backwards and forwards. But space-wise, we'll move things away from you in such a fashion that you can always only move forward.

Of course if we were born into such a world, we would all be used to it by now and it would seem perfectly natural. But in such a world, all of us would then be under the illusion that the space behind us was no longer in existence (anywhere) simply because *we* couldn't get to it.

So it is with time and us.

We can only move forward, time-wise, and so we naively and irrationally assume that what is in what we call the "past" is no longer in existence (anywhere), when the truth is only that *we* no longer have any access to it.

The Imaginary Egg Example

Get an egg, a whole, unbroken, uncooked hen's egg. Invite some friends over to your place, or, better yet, invite yourself over to some friend's place. [If you have no friends, put the book down and get some.]

Ask them to agree that this is a real egg. Don't let them do anything else to it, other than to touch it and agree that it real, that it does exist, and that it is, indeed, a whole, unbroken, uncooked hen's egg.

Take a picture of the egg with a digital camera, print it out, and have everyone sign their initials on the back, acknowledging the reality, that the photo represents, of the existence of the whole, unbroken, uncooked hen's egg. Write the time and location of the egg on back of the photo.

Get everyone, including yourself, to sign a piece of paper stating that you all agree that there is a real, whole, unbroken and uncooked hen's egg which really does exist. On this paper you all acknowledge that the photo is a visual representation of that egg (a representation of a *real* thing).

Now take the egg and smash it against the wall, grinning hideously, as the yolk slowly drips down someone *else's* wall (thus, the yolk's on them).

Ask your friends the following question: "Where is the whole, unbroken, uncooked hen's egg?"

They'll look at you as if light travels through your ears. That's all right, because they just won't understand. Don't let that be an excuse for them to give you the wrong answer. They will say something silly, such as: "It's obviously dripping down the wall, you moron!" (Of course, they'll probably insist that you clean it up before leaving their home.)

That would have been the right answer to the question that you did *not* ask, "What has *become* of the whole, unbroken, raw hen's egg?" It is not, however, the correct answer to the question: "Where *is* that whole, unbroken, uncooked hen's egg?" If they say, "It *isn't* any longer," tell them, "you mean it isn't located *here* any longer."

You see, even if we know better, on some level, we tend to think of things as *only* being real—*only* when they are in the present. Yet with each new, non-definable and constantly changing "now," everything that had always been "real" before seemingly becomes not only *un*real, but also totally *non-existent*.[5] We are forced to be hypocrites if we want to keep changing our minds about what is and isn't "now" and therefore what is and isn't real. In other words, we pretend that things, as they were, are not *still*, as they were, where they were, *when* they were.

This false belief, constantly re-enforced all the time, is partly due to our use of language.[6] It is also partly due to the fact that we habitually think of things as being located only *spatially*. That is because we have fewer restrictions regarding our use of space than we have regarding our use of time (in which we can travel in only one direction — forward).

Naturally, there is nothing located in space that isn't also located in time.

---

5. How fickle of us!

6. Language influences ideas at least as much as ideas influence language.

So the correct answer to our question is: "The whole, raw, unbroken hen's egg *is* (still) where it has always been." It remains in the same part of the universe, time-wise and space-wise, that it was in when you were with it. [We are, of course, not referring to the entire existence of the egg, or we'd be discussing the entire existence of the universe. We are not even referring to the entire existence of the life of the egg, as what we would all recognize to be an egg. Instead, we are talking about that piece of the existence of the egg when it was acknowledged to be a whole, uncooked, unbroken hen's egg — in your friend's kitchen, when you took its photo.]

Trick answer? No. Trick question? No.

Things can't be removed from the universe. They can't be "created," and they can't be destroyed.

Just because something changes[7] over a period of time and through space (or, over a period of *space-time*), does not mean that it disappears. It does not mean that it isn't still located *where* it always was, *when* it always was, in the various form(s) by which it is familiar to us.

An example of this is our whole, unbroken, raw hen's egg. No one is questioning the fact that it has changed over time (and space). The question is, *where* is *the egg that was* not *broken*? And the answer is written on the back of the picture signed by everyone who agreed upon the reality of the egg's existence (as a *whole, uncooked, unbroken* hen's egg). The location of that egg is in this universe[8], in the space and time that has already been documented (in this thought experiment at least, if not also for real).

Such an egg, as described, would not be any less real because *you* have moved away from it time-wise, than it would be if you had moved away from it space-wise. It's still there (and so are you—the you that agreed that it was a whole, unbroken egg). But you are changing your perception of "now" constantly,

_____

7. Everything changes (evolves) over time and through space. This is what I call *universal evolution* It's not only life forms that evolve, but *everything* within the universe evolves, or changes. Nothing ever stays the same for very long (and, as we know, nothing doesn't exist).

8. As if there were any others. . . .

which influences your notion of what is (currently) real.

Another way of explaining the same phenomenon, for those who don't yet understand it, is the following:

Is the 12th of July 1997, in the past, present or future? It certainly is in the universe—the same universe we are in now, the same universe we were in the last time we said it was "now," and the same universe we'll be in the next time we say it is "now". But is *that* day in the past, present or future?

You know that any event occurring on the 12th of July 1997 is a definite "event." You know that the content of that day (that is, whatever happened during that day in whichever location you care to name) was something very definite, specific, precise, finite and exact.

You know that the 12th July 1997 is in the universe, located in a time and space that doesn't change—just as the events that occur on that day also do not change. (For instance, the 22nd of November 1963 *is* when President John Kennedy was killed. [Notice that I didn't say "*was* when he was killed."] That fact remains the same always. We won't wake up one day and discover that his assassination really occurred on the 10th of May, 1876; and then on another day realize that he was actually murdered on the 29th of January, 2029!) Therefore, whatever occurred[9] on that day is what forever occupies the time and space defining that event.

Nothing about that day (or any other) will ever be anything other than what it always is. And yet we'll pretend that *it* is in the past, or in the present, or in the future, based only on where *we* are, time-wise, when asked that question.

If on or after the 13th July, 1997, you were asked whether the 12th of July, 1997 is in the past, present or future, you'd probably say (unless you wanted to give me a hard time for no good reason) that it's "in the past." If asked the same exact question (about the same exact date) on the 12th of July 1997 itself, you'd

---

9. One could just as reasonably use the word "occurs" in the present tense; in any case, whichever tense is used has the tendency to prejudice the reader into thinking that the event under discussion either is not still in existence, is not yet in existence, or is still currently happening for us. There's no getting around the linguistic re-enforcement of these false assumptions.

54

say, "It's in the present". And if you were asked that question on the 11th July 1997, or any time before then, you'd have said that the 12th is in "the future."

Yet the events of that date remain where they always are space-wise as well as time-wise. Nothing has changed in any respect regarding the event we call the 12th of July 1997. It remains the same. It's the same thing to which we are referring. So whether you claim the 12th of July 1997 (or any other day), is in the past, present or future, depends upon where *you* are when you are asked that question.

It is you who, moving through time, make contradictory remarks regarding that (or any other) date being in the past, present or future. It clearly depends upon where you are in the universe as to what you will claim about whether any particular date is in the past, present or future. But nothing about that event (date) ever changes in any way; it always remains what it is, where it is.

It is an event that remains forever in its little portion of the universe waiting for you to make up your mind about whether it has already occurred, is currently occurring, or has yet to occur.

The same thing can be said of any other date (after all, no day is any more or less real than any other), and for the same reasons. Therefore, one can understand that the future is already in existence in the universe — it's just that you haven't gotten there yet (in terms of your conscious awareness). Of course I have even more evidence that this is the case. But first . . .

Everything real exists. Nothing doesn't.

As we have seen, there is no such thing as nothing (no thing). There is no such place as nowhere (no where). What there is, however, is just the opposite. There is everything and there is everywhere. The universe is everything located everywhere. And the best part about it is that there can be absolutely no disputing the existence of the universe. (We'll deal with the solipsists later.)

Just as it is true that nothing doesn't exist and nowhere doesn't exist, so it is true that everything (real) and everywhere (real) does.

This is not to say that everything we can imagine exists in reality outside of our imagination, but rather, everything which is real is always really in existence. There is a limit on everything, though. There are a definite amount of things and places for those things. This is proven in the chapter dealing with the flaws of mathematics.

How Long Is "Now"?

When someone talks about what is happening "now," they refer to things that they know are real. You are reading this book. It is real. You are real. You are really reading it, now. But we also know that every time someone refers to "now," that it is a different event, occurring in a different time (and space) from every other "now."

So I ask you, how long is "now"?

Depending on one's point of view, the answer might be "Now is very short. It is infinitely short, and it keeps changing." Aside from the problem with the concept of something being infinite in any way whatsoever, there is the obvious contradiction here. That is, if "now" is so short a period of time, then how come it is *always* now?

"Now" is really not so short, after all. For every single individual, his or her idea of what is happening "now," is always current. They are always aware of "now." [Technically, of course, everything of which we are aware has already occurred in our past, since it takes some time for our sensory organs to transmit their information to our brain and for our brain to process this information into our conscious awareness.] The fact that we are forever changing what we think of as being in the present is nothing more than a psychological consequence of the human condition. For each and every one of us, "now" begins when we first become consciously aware (sometime before birth) until our death – with occasional interruptions during those times of sleep when we are not consciously aware, or if we've had a bit too much to drink.

Psychologically, that part of our brain that we call the "mind" is busily ordering the events of our lives into convenient and easily accessible segments of time and sequence. Yet over large periods of time, for most people, these memories of what has occurred become gradually more and more faint as they become more and more distant, and we can't be quite as sure of the contents of the events of our lives with the same precision with which we can be certain of what occurs in the now.

Therefore, we can see that since no event in the universe is any more real than any other event, those events to which we are attributing the notion of occurring "now," are somewhat illusory – especially if we pretend that they are any less real when they are either in what we call the "past," or what we call the "future."

The *past, present,* and *future* are all relative terms. There is no such thing as anything being *absolutely* in the past, present or future (for all coordinate systems, or frames of reference). This should be obvious to anyone who has read this book up to this point. To those for whom it is not obvious, keep reading, I have many different ways to show the same things to be true.

Time Is Curved—An Additional Interpretation Of Einstein:

Albert Einstein discovered certain facts about the laws of nature that he called his *Theory of Relativity*. This is divided into two parts: *Special Relativity* (which deals with frames of reference moving in an un-accelerated "straight" line[10]) and *General Relativity* (which expands this notion to cover acceleration, centrifugal force, and gravity).

I wish to explain some of this in my own words, and then expound upon it. The following material is true and has been verified by test and observation over and over again. [I say that because too many people think that when something

10. What is a "straight line" to one observer, can be curved line to another (in a different frame of reference) — and both observers' claims could be equally legitimate. Among many other things, Einstein proved that there is no such thing as an "absolute" straight line (for the above mentioned reason).

is labeled as a "theory," that it's just speculation, and therefore not necessarily true. When a theory has been tested and validated, it still remains a theory. That's what scientists call it, even when they know it to be valid. Sometimes this is called a Law, also.] The *Theory of Relativity* is not *just* a theory. It is also reality. The following is Einstein's relativity theory, summarized:

Einstein's Theory Of Relativity

Light travels at approx. 186,200 miles per second—for every observer, no matter what his frame of reference—provided that the light is traveling through a vacuum (nearly "empty" space, as in outer space), when unaffected by a gravitational field.[11]

Imagine that there are three observers, one of whom is traveling *towards* the source of the light (such as a star), one of whom is traveling away from the light, the last of whom is standing still with respect to the source of the light. For all observers, this same light comes to them at the same rate (the *constant* speed of light). This *absolute*, discovered in the laws of nature, is the speed of light (only when traveling through space where unaffected by gravity).

Imagine how odd this would be if we replaced the word "light" with the word "car."

Three people stand together. A mile away, a car is moving towards the three at sixty miles per hour. Person number one moves *towards* the car, also at sixty miles an hour. For him, the situation is the same as if he had stood still and the car traveled towards him at a hundred and twenty miles per hour. [That is, one could add the two velocities (the car and the man's). And in thirty seconds, the two of them would meet and become quickly and fatally acquainted.]

Person number two is actually *standing still* with respect to the sixty mile-an-hour moving car. [The car will greet him in one minute.]

---

11. Except if the observer were light itself, which, unlike any other observer, can and does travel at light speed (the absolute speed limit of the universe). If light did have a point of view, it would not be traveling at all — more on that later!

Person number three is moving *away from* the car at sixty miles per hour (the same rate as the car is moving, and in the same direction). That's as if the car were standing still with respect to that person (and vice-versa). [At that rate, the person and the car will never meet and they and their insurance companies will live happily ever after.]

What's true (on the surface) for cars is *not* true with respect to light. That is, you can add velocities in the real world on Earth, in most instances, without needing to take into account relativity.

You can't do that with speeds approaching that of light.

Light, unlike automobiles, comes to all observers at the same rate! What does this mean? Well, distance equals rate times time. And the speed of light remains the same for all observers, even as they approach the source of the light as fast as they can (for instance, at 100,000 miles per second—more than half the speed of light). So something *else* must change—that something else is *distance* and *time*.

And that, in fact, is what happens.

It seems strange to say that the faster you travel, the slower your time is. But think of it this way: the faster you travel through *space*, the slower you travel through *time*. Your wristwatch would slow down, as would your aging process and anything else that measures time.

But this would not happen for you from your point of view.

Things would seem the same to you, although from the point of view of an observer not traveling as fast as you, your time would slow down. That is, the rate at which you measure an hour, for example, would be slower than the rate at which I measure an hour (according to *my* clock), if, for instance, you were traveling near the speed of light and I were at home reading Kafka. Although, for each of us, one hour would seem to be the same "length" of time as it had always been. Your watch and mine would not agree with each other and therefore would not be able to be synchronized. You'd claim that my watch was running far too fast, and I'd claim yours was way too slow. And we'd both be absolutely correct. Time, the measurement of motion, is absolutely relative.

As you approached the speed of light, your time would slow down (from my point of view). If you could reach the speed of light (which you can't),[12] time would stand still for you. There's an absolute speed limit beyond which you can travel no faster no matter what you do.

Why not?

You need energy to accelerate. Acceleration makes you more massive ("G" forces applying to astronauts as they accelerate into outer space, for example.) The more massive you are the more energy is required to move you further. [It requires more energy to move a two-ton truck at the rate of 20 miles per hour, than it does to move a human being at that rate.] But the more you accelerate, the more massive you become—requiring more energy to move you any faster. So you eventually reach a limit, at which point, you require more energy to move you any further because the more energy you get, the more massive you become—requiring even *more* energy to move you any further. Eventually, you reach a limit beyond which all the energy you get no longer goes into moving you any faster; it just makes you more massive requiring *more energy* to move you any faster. In other words, you reach an absolute speed limit (which may be close to, but will always be less than, the speed of light).

At the speed of light, you would become *"infinitely massive"* which, of course, is an impossibility.[13] So you can't move that fast, *ever*. And you certainly can't move any faster—either scenario would have logically contradictory consequences.[14]

---

12. As we know, time waits for no one. [Death, on the other hand, grows more impatient daily.]

13. There are several reasons why "infinitely massive" is not possible; for one thing, the concept of infinity itself is invalid (to be proven later in this book) which is reason enough; for another, if there were such a thing it would mean that the mass of the object would be infinite and therefore never-ending. If it were never ending, it would have to be taking up infinite space – space which is already being occupied by other mass and, as we know, two or more things cannot occupy the same space at the same time [though it wouldn't be the *same* space if it weren't at the same time, anyway]!

14. Traveling at the speed of light, your time would stop completely – going any faster than that would mean you were traveling backwards in time which, as will shortly be pointed out, can never be possible (and, in fact, makes no sense).

But *light* can move that fast. In fact, it must! The only situation in which light travels any *slower* than usual, is when it travels through some medium other than "empty" space—such as through water, or when affected by a gravitational field (as is produced by the Sun, or by a black hole).

Light, which behaves as waves sometimes and as particles at other times, is actually made of particles called *photons* (named by Einstein, who posited their existence). Photons can be affected by a strong enough gravitational field. Proof of this theory came during the total eclipse of the Sun, when the planet Mercury was in alignment with the Sun.

Normally, we would not be able to see where Mercury was in the daytime sky, when the Sun's light would make the much fainter light of Mercury impossible to detect. During a total solar eclipse, however, the Sun's light is covered completely by the Moon. Under those conditions, for a short period of time, we can see Mercury (when the two are aligned properly).

Einstein predicted that the light traveling from Mercury to the Earth would be pulled away, or bent, slightly towards the Sun, due to its gravity (thus the notion of "curved space"). This would make Mercury appear (very slightly) to be somewhere else in the sky, other than where astronomers knew it to be. The normal calculations, used to find Mercury, were based on Isaac Newton's mechanical description of the universe. Einstein took into account the effects of relativity. As a result, *his* prediction of where Mercury would be found varied a little bit, but still measurably, from those of Newton.

The scientific community wanted to test to see which of the two physicists was more nearly accurate in their prediction. If Einstein were correct, then a lot of what Isaac Newton had taken for granted (that time and space were absolutes—the same quantities of each would always be the same throughout the universe for all observers at all times) would be proven false. If Newton were correct, however, Einstein's theory would fall apart.

Einstein was proven right during a solar eclipse in the 1930s, when physicists on a specific expedition in Africa verified his predictions, just for that purpose. Einstein was not the least bit surprised. He knew that if the results hadn't verified his theory, it would have meant that the scientists had made a mistake.

Einstein predicted that the gravity of a massive-enough object would curve the *space* around it, causing whatever had to travel in its path to follow this "curved space." Since space has no form, it confuses many people when it's described as being *curved*. All that really means, however, is that when something in outer space (light, for example) travels from point *A* to point *B*, it always does so as economically as possible. That is, it takes the *shortest distance*—always. And the shortest distance between two points *isn't* always a straight line. Sometimes it's a curved line (when the space is bent by a massive enough object). Since you can't see curved space, you'd only know it was curved by following the path of whatever was traveling between the two points.

Contrary to popular belief, it isn't really space that is curved, but the path that objects (including light) must take to get from one place to another when traveling through space—when affected by a heavy-enough gravitational field (the pull of gravity from an enormous black hole,[15] for example).

Einstein proved that energy and matter are essentially the same thing (E = MC^2), thereby re-naming their combination as "energy-matter." He did the same thing for the dimensions of space and time, by combining them into one concept. Nothing can exist in time without also occupying space. And nothing can exist in space without also occupying time—thus, "space-time." And just as it is true that energy and matter are two sides of the same coin, so it is true for time and space. Therefore, what affects space also affects time.

If time and space are forever inter-connected (and they are), and if gravity curves space (and it does), then it follows logically that gravity also curves time. Physicists haven't seemed to figure this out yet, to the best of my knowledge;

---

15. A Black Hole is just a star, like the Sun (only perhaps a million times larger), which has burned enough of its energy from its interior to cause the gravity from the star to pull in on itself, thus collapsing it into a drastically smaller entity with the same gravitational force. It's a pull so strong that everything is attracted to it, even light. So light can't escape from it, and therefore we can't see it, thus the term *black hole*.

or if they have, I am not aware of it. Perhaps they don't know how to *interpret* the idea that gravity *curves* time, therefore, they would incorrectly assume that such a concept was meaningless. Because if they did know how to interpret it, they'd not continue claiming the Big Bang was the "beginning" of anything – let alone of the universe of which the Big Bang was merely a part. So claiming the "Big Bang" to be "the beginning of the universe" is as irrational a statement as saying that someone could be his own father, or his own son.

Then what does it mean to say that "time is curved?"

It means that if we could travel far enough into the future, we'd eventually be back in the past. [This is precisely why terms like "future" and "past" are relative and not absolute. "Ah," you may tell me, "but you said we always travel into the future and cannot go back in time." To which, I would reply, "Yes, that's still true for all material objects including ourselves."] Unlike other (spatial) dimensions, time moves in one direction only: *forward* (that is, from what *we* call the "past," into what we call the "future"). From our point of view, we are *always* moving from the past and into the future. There seems to be no question at all regarding what "is" in the past and what "is" in the future. If we were able to live forever, surviving the perpetual Big Crunch and Big Bang continuum,[16] this would no longer be quite so obvious.

Imagine that our voyage through time were like riding in a train on an unbelievably vast, extremely long, rail line. Imagine the train track *not* to be a straight line, moving forever in one forward direction. Instead, imagine the train to be on a circular track so large, so vast, that it would seem like a straight line in both directions unless you could live forever (something that, while obviously impossible, is not unimaginable).

Imagine riding through all of existence—all of the universe itself.

---

16. I know I haven't explained the *Big Bang/Big Crunch* continuum yet; be patient, it's coming. But for purposes of continuity, it would only be distracting to discuss it here and now on a tangent – even more distracting than these damned footnotes!

Eventually, you'd come back to where you "started" from (that is, to wherever it was where you got on).

Now imagine that this gigantic train ride is going through only a tiny part of the universe, on a tour of *your* entire life. Everything ahead of you (that is, in front of the train) is in your future; and everything behind you (in back of the train) is in your past.

At this point it seems absolutely obvious as to what *is* in the future and what *is* in the past. [The real reason for this illusion is due to our everyday experiences, which make it somewhat difficult to analyze this objectively.]

But suppose we extended your train ride in part of the universe to allow you to view things beyond your tiny life span? [After all, if you take the amount of time you are alive, and divide it by the amount of time you are not, you'll have a very small number indeed!]

Suppose we allow you to live forever. What would you see then? What would your view of the *past, present* and *future* be like under those circumstances?

It is my contention that since what affects space equally affects time, and since gravity curves space, it must also curve time.

The Big Bang/Big Crunch Continuum

The universe as a whole is made up of the sum of its parts. The content of the universe is made up of energy-matter (even the nearly "empty" vacuum of outer space cannot exist unless bounded by objects which themselves are made of energy-matter). Neither energy nor matter (energy-matter) can be "created" nor destroyed. If you add up the contents of the universe, you get the universe as a whole. Therefore, since everything that makes up the universe cannot have been created and cannot be destroyed, it follows that the universe, as a whole, could not have been created and cannot be destroyed.

The *Big Bang*, therefore, is that part of the universe during which time and space exploded into the expanding part of the universe in which we currently find ourselves. Since nature always takes the easy way out, always accounts for itself, and always maintains its balance, it follows that:

The expanding stars and galaxies will run out of energy. At that point, the gravitational forces (most of which are virtually undetectable—in the form of dark matter, such as black holes) will pull the stars back in on themselves— eventually forming the *Big Crunch*. At such a point, the *Big Crunch* will lead directly back into the *Big Bang*, forming a *Big Bang/Big Crunch Continuum*. And it will be the same *Big Bang*, because it is located *time-wise* as well as space-wise, and *space-time is curved!*

The gravity of the entire condensed mass of the *Big Crunch* (which would make the gravity of a black hole seem trivial in comparison) curves the space-time of the *Big Bang/Big Crunch* continuum. And, in fact, the near-singularity "point" into which the *Big Crunch* condenses, is the exact same near-singularity "point" that *is* the *Big Bang*.

In other words: The "Big Bang" is that part of the universe into which the "Big Crunch" implodes. Therefore, all of time itself is *finite*, yet unbounded, in the same way that the curved line of a circle is finite but has no bounds.

I realize that many people will consider this to be an interesting theory, but will not understand it to be valid until they see some independent verification. [They don't realize that the evidence is already here and the correct interpretation of this evidence is logically sound and needs no additional empirical confirmation.] Anyway, perhaps one such way such people can verify my theory would be to discover blue-shifted stars older than the alleged age of the universe itself (that is, older than the age of the universe as measured from the moment of the Big Bang).[17]

---

17. The universe has no real age, of course, and it's absurd to claim that it does. But if you pre-tend that the *Big-Bang* is the "beginning" of the universe, then eventually some device, such as the space telescope, should be able to detect blue-shifted stars (that is, stars moving *towards* us) that are older than a universe whose "age" is measured from the *Big-Bang*.

More Evidence That the Universe Has Always Been Here,

And That The Future Is Already In Existence:

It still may not be obvious, to some of you, that not only is the future completely determined, but it is already in existence. Previous explanations regarding this have included the idea that all events are equally real, including all "now" occurrences, no matter whether they've already happened or are yet to occur, and that all events are located in time and space permanently (in the universe). That means that everything that happens stays where it is and can never be gotten rid of. This is true, in spite of the fact that our ever-changing current awareness of what is *now* gives us the illusion that what *was*, is no longer. However, since this is a foreign idea to most people, and since it requires some counter-intuitive thinking, I am going to present yet another way of demonstrating this to be true. But first:

The Universe Must Be Considered in its Entirety and Cannot Have Been Caused.

We remember that it's always "now" for everyone – all the time. Even though the now changes with every other "now," and even though we keep contradicting ourselves and each other regarding what it is that we call the present (or the "now"), it can still be understood to be a length of time which has a permanent and always current existence.

One of the consequences of this thought, is an understanding that there is such a thing as a point of view of the universe as a whole, in its entirety, which is frozen forever in space and time. How is this?

Everything that occurs *within* the universe (except for light—which we'll analyze shortly), which *is everything* (other than the universe taken as a whole), does so because of the universal law of cause and effect. That is, things are the exact way they are at any given moment because of the way they were the moment before. Things will be the exact way they will be a moment from now because of the way they are now. [This is not a radical idea and it's certainly not difficult in the least to understand. For example, when one travels from point A to point B, one gets there by traveling in between the two points in a progressive fashion; that is, you don't find yourself very near to point B and then suddenly being closer to point A again! There may be an apparent contradiction to this in the Quantum world, but that will be dealt with later.]

If one understands causality within the universe, then it follows logically that by extension in both directions (into the "past" and into the "future") we live in a completely determined universe where everything happens the only way(s) it can. This has previously been shown in an earlier essay.

What has yet to be resolved is the age-old apparent paradox[1] regarding the seemingly infinite extension of causes in both directions. That there would be such an extension in either direction, let alone in both, is enough to be very disconcerting indeed. But just because such an explanation is not what you want to hear, doesn't make it false. This one, however, conveniently does happen to be false anyway.

Like the mighty mythical Atlas, supporting the heavens on his shoulders, the question always comes down to the feeble concept of a first cause—such as who, or what, is supporting Atlas, etc., ad infinitum. [Another proof that the universe has always been here, and therefore could not have been caused, comes from the fact that infinity is, you guessed it, yet *another* illusion. Since there can be no infinity (to be explained later in the book), nothing can be infinite. Therefore, nothing (which has an actual existence apart from the world of mathematics) continues forever without eventually repeating itself — in any direction — neither time-wise, nor space-wise.] The universe as a whole (and it *isn't* the universe *unless* it's considered as a whole) cannot have been caused for

1. All paradoxes are only apparent, as they can be resolved logically or shown to be meaningless or based on a false premise.

several reasons already mentioned previously, as well as the one about infinity, to be dealt with later in this very book.

Here's another reason: causality implies that one event or action influences the outcome of another event or action. All causes are also effects (and vice-versa). But all causes absolutely require the passing of time. One event can't influence another unless an action of some sort takes place. And all actions of all kinds require time during which to occur.

At the speed of light, time stands still (but only for that which is traveling at the speed of light, namely, light)! There can be no motion for light, from its own point of view, as it is frozen in time forever. And light travels throughout the entire length (both space length and *time* length) of the universe; therefore there is a point of view from which the universe as a whole can be considered. And in *this* point of view -- where the light which is, from our perspective, still ("currently") traveling in one direction and has yet to reach its destination (for example, a star some several thousand light years away) the light is clearly already where it, according to us, has yet to be.

In other words, it's already everywhere in its entire path all at once – its "past," "present," and "future." This necessarily means a universe in which everything is permanent and the future is already in existence.

Again, from our point of view, light takes time to travel (for us, and for everyone and everything *not* traveling at light speed, it's always roughly 186,200 miles per second unless slowed by gravity or by traveling through some medium other than a vacuum). It's very important to understand that for us, the event of some light from a distant star, say, 2,000 light years away, has taken two thousand years to travel from its source to reach us. But as Einstein has proven, time is frozen at the speed of light and from that point of view, it didn't take two thousand years at all, it took no time whatsoever, because it is everywhere it will ever be, all at once. That means, necessarily, that all the places where it "will" be, from our viewpoint, the light is already there from its perspective, which could only be true if the places (time-wise and space-wise) were also already "there," which, in turn, means all of the universe has to have always existed.

I realize this is not necessarily as obvious to some as it is to others, and yet I don't wish to insult my readers be dwelling on something they may already understand, so please forgive any unnecessary redundancies.

Remember, Einstein proved that to the outside observer, one's time slows down the faster one approaches the speed of light—to the point where it stops altogether, at that speed. So even though light is not a conscious entity, and therefore has no awareness, nevertheless, such a point of view (or frame of reference) does really exist, has true meaning and is useful for us to discuss. What does it mean to say the point of view (or, frame of reference) of light is useful for us to understand?

It means that we can look at the same event in two totally different ways, and even though they seem to be contradictory and mutually exclusive, they are not. They are both equally valid descriptions of the same real event. [The first one is ours: light has taken two thousand years to reach us from a star two thousand light years away. The second description is from the frame of reference of light, in which it took no time to get here because its time is standing still and therefore is always everywhere its entire path will take it (from our point of view) all at once.]

Another way of looking at this is to imagine the universe as if it were a feature film. Motion picture film reels are usually put into projectors in what are called "platters." Platters are flat spools onto which the entire feature film is placed after being threaded through the projector, such that the last frame of the film is attached to the first frame. In this way, after it has run through an entire showing of the feature, it is already ready to go again and doesn't need to be re-wound.

There is a (literal, in this case) frame of reference through which light from the projector's bulb shines through the film to project the "moving" image upon the screen. This is called the aperture. It is from this point of view that we see movement, cause and effect, taking place in a relatively small piece of space over a ninety-minute (for example) period of time. This is how we are used to seeing a film.

But we could view the whole thing all at once, if we were to take it off the platter's spool and stand back far enough from it to see the entire feature film appear as one long strip—one entity frozen in time, covering a much larger piece of space than is covered by only the aperture. And we would clearly recognize that these two different, seemingly contradictory and mutually exclusive points of view were of the same exact thing. We'd know that they were equally valid and that from one viewpoint, there is time and motion (and therefore causality). But from the other one, that of the unraveled film as a whole long strip, there are all of the film's events forever frozen in time permanently and no motion.

So it is with the universe.

We are within it, traveling at speeds always less than that of light. For us (and for every event) there is time (the measurement of motion) and therefore there is motion and therefore there is causality.

For light, we are infants along one part of its path, adults at another, and dead for billions of years at yet another—all at once.

For the universe, it is forever frozen, permanently, containing all of (its) space and all of the time under which we, less-than-the-speed-of-light travelers, operate.

Imagine the universe as that reel of film. In this analogy, the film reel would be very, very long indeed. So long, in fact, that the distance from the (arbitrarily chosen) first frame of the film to the last frame of the film would be separated by perhaps billions of light years.

[This is not such an odd thing to imagine, is it? After all, if we were to watch the unfolding of all the events within the universe from our current coordinate system (frame of reference) until the time when things started repeating themselves, the actions, or events, within the universe would be unfolding over quite an extremely long period of time. And so, instead of looking at everything in a relatively small distance of space over an extremely long period of time, we are looking at it all "at once" over a very, very long distance of space. But we could not see the entire universe/film reel all "at once" immediately, because the distances between the first and last frames would be so far apart, that it would take billions of years sitting in the same spot, the film not moving at all relative to us, for the light from all of the frames of the film to finally reach us.]

The universe *is* the sum of its parts. Those parts include totally different pieces of time which would be truly contradictory were they to be applied to the entirety of the universe as a whole in the same way in which they are applied to the entities within (such as ourselves). For the universe itself, there is no time in an overall sense; that is, space-time is frozen forever.

Time is not absolute, anyway, as Einstein proved.

That means, necessarily, that the universe (as a whole—don't forget it has to be the entire universe with all of time and space or else it isn't the universe at all, but only part of it) must be considered as being frozen in time. And that means that there can be no question of causality regarding the universe itself, because it is meaningless to impose causality on anything that is not subject to the passage of time (such as light, and the universe itself).

From our point of view, there is motion through space and therefore there is time and therefore causality. Our point of view is equally as valid as the point of view of the universe (and of light); so how can these apparent contradictions be reconciled?

Easily.

There is no (absolute) beginning to the universe (nor to anything within it). There is no absolute end, either. This is true of the rim of a circle. Like our friend, the universe, a circle's rim is finite yet unbounded. That is, there is a *definite* amount of it such that if you were to continue around long enough, you get back to where you (arbitrarily) started from, and would then repeat your same journey.

Applied to the universe, if we wish to label the near "singularity,"[2] from which exploded the *Big Bang*, as being the (arbitrarily proclaimed) "beginning" of the universe, then we are to understand that the cause of the explosion of this

---

2. One of the consequences of Einstein's *Theory of Relativity*, is that the *Big-Bang* exploded from a "singularity," which is a point. The problem with this, is inherent in mathematics itself (to be dealt with later). For now, I'll just say that points have no dimension, therefore they don't exist. The *Big Bang* exploded from something larger than a non-dimensional "singularity."

"singularity's" *Big Bang*, is the implosion of the "singularity" of the *Big Crunch* which "singularity" would be identical to the *Big Bang* as it would have curved space-time as far as possible, back in on itself, and would therefore be located in the same exact position in space-time within the universe.

In other words, the "end" is attached to the "beginning" just as is true with our example of the feature film platter. Or, if you prefer, the *Big Crunch* caused the *Big Bang*.

What's 'outside' of the universe, and what does the universe look like?

These are two favorite questions of many people. What's outside of the universe, and what does the universe look like? Again, we are faced with the same problem with these questions that we are faced with on so many others. That is, the very questions themselves take things for granted which aren't true – based on false assumptions.

False assumption number one: the concept of the "outside" of the Universe has any real meaning. It does not. The very reason we imagine that it might have real meaning is because we live in dwellings that are covered by a roof. Early man lived in caves. When something encloses us in a little box of some sort, we are "inside." Remove ourselves from that box, and we are "outside." Look up and we see the sky. It, the stars, the sky, is what we have always called "outside" (that is, it is outside of our frame of reference, or coordinate system, which we call the Earth)!

And yet *it* (the stars, the sky) clearly is inside of -- that is, simply, within, the universe -- as are we, as is everything else. The universe contains all coordinate systems, all frames of reference, all planets and stars, all "empty" space. The universe itself cannot be considered as a frame of reference, nor as a coordinate system except when compared to one within the universe. In other words, the "outside" of the universe as a whole entity – and it isn't really the universe if it isn't the whole of all events which ever occur – is everything within the universe itself. That is, the universe is both inside and outside at the same time. Therefore, there is no meaning whatsoever in claiming that there could be an outside to the universe which isn't also at the same time the very inside of the universe.

The outside is the inside.

The next question, "What does the universe look like?" rests on the false premise that it can "look" like anything at all. This is a little tricky and deceptive. Einstein demonstrated how to imagine what the four dimensional universe looks like, by analogy. He compared a two-dimensional world to a three-dimensional one, and then did the same thing for our apparently three-dimensional universe to help people visualize a four dimensional one. But that is illusory. The problems with that are as follows: To imagine what something looks like, one must be able to imagine seeing it all at once. That is, when we think of an object such as a pencil or a chair, we can only do so because we can imagine seeing all of the components of the object (even if only from one particular viewpoint) all at the same time. But the entire universe includes time, and what does time look like? Also, if we were to think merely of the material contents of the universe as a whole, then there is no point of view from which anyone would be able to see everything all at the same time, since the objects of the universe are separated by such vast distances that it is never the case that we would be able to see even a large percentage of every object "at the same time," much less see all of the objects at once. Finally, in order to imagine anything in its entirety we must have an image of it against something else – that is, against some *other* background.

For example, try imagining just a pencil without anything else to compare it with (including "empty" space or even some neutral color background). It is natural even to imagine individual objects rotating in space so as to see these objects from various angles. Take away the space and leave only the object remaining, and our imaginary image of these objects becomes even less clear.

There are many objects in the universe that we cannot see – not because they aren't there, but because their placement in time prevents their image from ever reaching us. Anything located ten light years away from us "now" will appear as it was ten years ago (from our measurement of time). So whatever may be occurring there "now" (again, from our measurement of time) will not be visible to us for another ten years, since that's how long it would take light from that object to reach us.

Yet the entire universe contains objects separated by astronomically vast regions of space-time such that there is no image of the entire universe to be imagined. It, too, is an illusion and claiming that it (a 'view' of the entire universe) exists such that all of its contents could ever be seen at once to any observer within the universe, is claiming something untrue. So even though there is such a thing as an imaginary point of view of the universe as a whole, it is a point of view in which the entire universe has no clear "look" to it.

The Twin Earth Future (Apparent) Paradox

Here is another thought-experiment:

Imagine that there were a twin planet to Earth, with the same kind of calendar and system of keeping time as we have. Now imagine this planet is cleverly called *planet X*.

It is always exactly ten light years[3] away from Earth. Let's also imagine that *planet X*, like the Moon, always has the same side facing Earth. While we're at it, let's endow you with incredible eyes (perhaps you already have incredible eyes)—such that you can see the year on a coin resting on the surface of *planet X*.

You look outside your window and see your "twin" apparently waving hello to you. What you see "now," your twin has done ten years ago (since that planet is ten light years away and it has taken the light from that event a full ten years to reach you "now").

---

3. A *Light Year*, is the measurement of the distance that light travels in a (non leap-year) year. Light travels at approximately 186,200 miles per second. It travels eleven million, one hundred seventy-two thousand miles in only one minute. There are 1,440 minutes in a single day. So you can just imagine how much distance light travels in an entire year.

You wave back. Your very own twin (whom we shall call "Twinny") won't see that greeting for another ten years (because that's how long it will take the light from *the event of your waving back* to reach Twinny on *planet X*). Therefore, you are "now" ten years into Twinny's future—and everything you do for the next ten years will be in your twin's future when it becomes part of *your* past.

Yet both of you are "currently" occupying the same universe right "now." Similarly, Twinny is also in existence now on *planet X* where you can't detect anything your double does for the next ten years, because Twinny is ten years in *your* future, right *now*.

At least in that sense, we can see how the future is already in existence (for each of you, from the point of view of the other). Of course, there is no such place as *planet X*, ten light years away from, and parallel to, Earth.

But there is such a region of space-time. And there is also such a place two, three, five, twenty, one hundred thousand light years away. Furthermore, all we can ever see is in the "past." The further away something is in space, the farther back it is in time.

Yet we exist here and "now" forever in the future of anyone far enough out in space to be able to receive the light from this (or any other "current") event. And vice versa for us. But what about the future here on Earth? You may be wondering how future events on Earth can already be understood to be in existence. After all, we're not located any amount of light years away from ourselves.

Well, it's really not all that different from the above-mentioned example. The light from the Earth of, say, 1972, is still traveling throughout the universe, yet nothing we could ever do would allow us to catch up with it. Why not? Because it's traveling away from the Earth (of 1972) so we would have to travel *faster* than light in order to catch up with it, let alone to surpass it. Naturally, you know that our time slows down the faster we travel through space—such that near the speed of light these effects can begin to be detected. [Note: this is a fact about light and time that Einstein discovered.] At the speed of light, which we can never reach, we'd be infinitely massive and our time would stop. We can't exist without time. (And if we were infinitely massive, we'd probably need to go on a severe universal diet immediately!) Therefore, the absolute speed limit for us is less than the speed of light. And since we can't ever reach light speed, we certainly can't surpass it. And since we can't surpass it, we can't *ever*, under *any* circumstances, re-locate ourselves somewhere else in the universe where we could look at the Earth's "past" from our "present" point of view.

The Imaginary Camera Example

Let's take a brief detour—a temporary tangent, if you will, on our trip around the universe. I promise to take you back home again and to continue from where we left off.

Let's imagine three old-fashioned 35mm still cameras that use film, like in the old days. Let's call them cameras A, B, and C (for purposes of clarity, not just because those names are so clever). There is a car moving quickly down the road at twilight, with its headlights on. This car is in front of your home. You take a photo of the car with each of the three cameras at nearly the same time.

Camera A has very high-speed film so that the shutter is open for only one tenth of a second. Camera B takes a picture with a shutter speed of one full second, while camera C's shutter remains open for a full ten seconds. Each of the cameras is taking a picture of the same thing (event). You get the photos developed. The photo taken from camera A looks like what you'd expect a photograph of a car to look like. The photo from camera B looks a little blurry, while camera C's picture shows a streak of the car's headlights and looks more like a special effects photo than what you'd expect from a picture of a car.

Now I ask you a simple question: Since all three photos were of the same event, which one is a more nearly accurate representation of the car? Most people, unless they figured this to be a trick question, would choose the photo from camera A, the one with the shutter speed of one tenth of a second, as being the best representation of what the car looks like. Yet, camera C had its shutter open for a full ten seconds—that's one hundred times longer than the shutter speed of camera A.

Now I'll ask you again: Which of the three photographs takes into account more information about the life of the car? That is, which photo *really* looks more like the car—not from our point of view, but from the point of view of the entire life of the car—from its arbitrary beginning (the day it "rolled off the assembly line"), to its arbitrary end (the day it was towed to the junk yard)? The answer, as if you hadn't guessed by now, is the photo taken from camera C, whose shutter was open for a full ten seconds. Sure, the car appeared to be a blurry streak of headlights. And obviously, ten seconds is a very tiny fraction of time in the life of something that exists for, perhaps, twenty years. But it's certainly longer than one tenth of a second. And what does anything look like if we wish to see the whole thing?

Even starting with the arbitrary beginning and ending (after all, nothing really ever has an absolute beginning nor ending), we think of things as looking like what they look like to us, in our current "now." That is, there is a period of time during which we are aware of the "now." Although it keeps changing, any given "now" still lasts for a certain definite (though very brief) period of time. So naturally we judge the "look of the car" compared to what a moving car looks like to us. We're seeing the same ten full seconds in the life of the car that camera C recorded—longer, even. But we perceive it in tiny increments; closer to the way camera A records the event. [In other words, we don't notice the space-time continuum.]

Now let's bring all of this back to our earlier example. Pretend the three cameras are hundreds of thousands of miles away, photographing the Earth as it travels through space. Let's add lots of cameras with progressively longer shutter exposures. Unlike ourselves, these cameras would not have the psychological (and/or physiological) restrictions that prejudice us to believe that what we see *really* "looks like" what we see.

These cameras would record progressively longer-exposed pictures of the Earth. What they would show us would be a blurred spiral Earth, coiling around the slightly bent pole of the Sun (which itself is constantly traveling through space).

In other words, a picture of the Earth, as taken from the point of view described above, even for a period of just one decade, would look much like what I've described, and little like anything we are familiar with. If we could map events taking place on the Earth in such a photo, we would see how any individual's life, like the life of a car, and of the Earth and Sun, etc., is a continuum.

We know that no event (occurrence in time and space) is any more or less real than any other event. We have also established that what "is" in the past, present or future has nothing to do with the event itself and everything to do with where we are, time-wise, when considering that question. We also know that the same exact event (for example, today) we have claimed, at different times, to be in the future, in the present, and, eventually, in the past.

Therefore, there is no absolute past, present or future. All events are always already in existence from the point of view of the universe as a whole (which is the point of view of light). And somewhere else in the universe is tomorrow's planet Earth with your entire day already in it. And the same is true for every day in your life, and every day of Earth, and every event of every nature everywhere throughout all of time and space.

In Summary:

The Universe is all of time and space and its contents (energy-matter existing in space-time). Neither energy nor matter can be 'created' nor 'destroyed.' Add up the contents of the universe, and you have the universe as a whole. Therefore it, too, cannot have been 'created' nor can it ever be 'destroyed.'

The Universe isn't the universe unless it is considered in its entirety as the whole of space-time and energy-matter. The point of view of the universe as a whole is equal to the point of view of everything within the universe, which is the same as the point of view of the path of all light. Time slows down as matter approaches the speed of light. At the speed of light, time stops. So from the point of view of both light itself --and therefore that of the universe in its entirety-- time stands still. When time stands still, there can be no causality (because cause and effect is a description of actions which require time during which to occur). Therefore, neither light, nor the universe as a whole, can be, or have been, "caused."

Therefore, the universe has always existed and is permanent. That means that absolutely everything (all events within the universe, which is all events, period) occupies its specific and permanent region of space-time which we think of as either being in "the past," "the present," or "the future," based only on where we are in time and space when considering this issue. Therefore, both all of what we claim to be "in the past," as well as all of what we claim will be "in the future," is always a permanent fixture within the universe.

For these, and other reasons, the *Big Bang* cannot be considered as the 'absolute' beginning of anything, much less the universe itself, of which the *Big Bang* is only a part.

The *Big Crunch* occurs in that region of space-time before which the energy of expansion of the *Big Bang* has/had been depleted, and the gravitational pull of the matter of that section of the universe begins/began to pull in on itself. In what amounts to both the conservation of mass-energy, as well as the most rational explanation of all the causal connections within the universe (for everything traveling at less than the speed of light), the *Big Crunch* is the implosion of all of the material of that region of space-time, into the near 'singularity' which is located in the same exact spot, and is therefore the same exact event, as that which we call the *Big Bang*.

The Folly Of A god And The Problem With Religion

Some people require no evidence in order to believe,
and an "infinite amount" (sic) in order not to.

"*Faith*: Belief without need of certain proof. [Synonymous with the word *religion*]"

"*Superstition*: A belief founded on irrational feelings, especially of fear, and marked by a trust in, or reverence for, charms, omens, signs, the supernatural, etc.; also, any rite or practice inspired by such belief."

"*Delusion*: A false, fixed belief, held in spite of evidence to the contrary." [1]

During all of the history of mankind, there has probably never been a force so despicable, so vile, cruel, hypocritical, politically manipulative, degrading, controlling, counter-productive and downright evil as the institution we foolishly cherish, called religion.

Nothing has had more of a detrimental, dangerous, irrational, destructive and stupefying effect on people—over the many thousands of years of its existence in various incarnations.

It is superstitious through and through. And while it is frequently satisfying to many people, the emotional dependence upon, and addiction to, this multibillion-dollar a year industry is often no less ruinous to the individual's personality than a similar addiction to a dangerous drug would be.

---

1. According to at least one version of *Funk and Wagnell's Standard College Dictionary.*

This is not a new idea, as Karl Marx pointed out when he claimed religion to be the opiate of the masses. But it is certainly true, and almost literally so.

When the adherents of a religion become numerous enough, that which would have otherwise been abhorred by some, and labeled as a cult by others (and seen by many to be harmful to its followers), then becomes a beloved religion, elevated to a status of bizarrely inflated false goodness and worth.

This is not to say that individual people who think of themselves as being religious are by any means automatically bad people – far from it. Many of them are among the best humankind has to offer: kind, generous, helpful and charitable. But way too frequently, they themselves are the victims of an institution whose purpose and function has far more often been to control and manipulate the masses than to help them.

Why?

Throughout all of our existence, humankind has wanted to understand our environment.

We looked upon the midnight skies and saw sparkling lights. We wanted to know what those twinkling little things were.

We also wanted to know why, during certain times, it would rain. Why, when it rains, do we sometimes hear real loud noises?

Why is there fire? Why wars? Why erotic sexual urges? Why does grape juice turn into wine?

To early humans, *everything* seemed mystical. The world was filled with magic, so that the only satisfying explanations would also have to be magical.

Most people were (and, unfortunately, still are) filled with fear. Some had curiosity. Others had both. The human race had just enough penetrating intellectual inquisitiveness to help separate us from most other animal species. This curiosity was absolutely necessary for any progress or advancement of society.

As humans slowly evolved, the masses (who, of necessity, are average in every way) came up with the simplest, most easily understood explanations for the various physical phenomena that they observed. The problem is, of course, that the simplest of explanations, while accessible to the simplest of minds, all too often accurately explains absolutely nothing.

On the other hand, a minority of other people wanted their explanations precisely to correspond with reality. They wanted to make sure that they understood the *cause and effect* relationships that exist between that which they had observed and the undiscovered laws governing those events.

So, over a long period of time, these two diverse groups worked separately to try to understand and explain the various mysteries that they had faced.

The masses were satisfied with the myths about creatures that were just like themselves (that is, in their own image), except that these beings had special magical powers. The males were called *gods*, while the females were called *goddesses*.

Each of these mythical characters came complete with its own individual and charming legend – guaranteed to please the populace.

But the true intellectuals were not satisfied with this popular "understanding" of natural events. They wanted their explanations to make sense and not to be prejudiced. Prejudice, by definition, is "pre-judging," that is, judging in advance of having the facts, or evidence, from which to form a logical and reasoned opinion.

Religion, by its very nature, is prejudiced.[2]

It makes bold and unsubstantiated proclamations about reality and tells its followers to ignore all evidence to the contrary (much like *The Wizard of Oz* told Dorothy to "pay no attention to the man behind the curtain").

Science, unlike religion, is not so arrogantly pretentious. People who are intellectually honest, use their natural curiosity to explore reality in an unbiased way. They might have an idea about how something works, in which

---

2. Prejudice: a judgment or opinion—especially an irrational one—formed in advance of having the facts, evidence and information.

case they may form a theory, but they won't claim it to be valid without proof. The theory must be shown to correspond with reality before it can be accepted. Were there to be evidence to the contrary, the theory would be changed (usually sharpened and clarified) or completely discarded if that's what the evidence called for.

The scientist, or any rational, honest intellectual, throws out his beliefs when faced with facts that contradict them. The religious follower, however, when confronted by facts contradicting *his* beliefs, throws out the facts. And then, like a good follower, smiles contentedly. It's as if he had just scored well on some mystical test of his blind faith. As if he had been rewarded for his adherence to whichever set of rigid religious dogma he had devoted himself.

This is what explains the divergent paths taken by science and religion when attempting to deal with similar subject matter (that is, with certain statements about reality).

Why, then, are some fairly well educated and apparently otherwise intelligent people also religious?

First of all, there's a higher percentage of non-believers among this group, than among the general population. [3] Secondly, very few such people, who *do* consider themselves to be religious, would claim to be orthodox followers of their religion. Thirdly, there is an unconscious psychological component that is completely independent of the intellectual abilities of the believer. And it is this factor, the unconscious emotional influence, which causes such people to believe as they do, in spite of, and certainly not because of, their intellect.

For example, suppose that a man becomes a priest at an early age, because his family made him attend church on a regular basis and, in general, placed a high value on religious beliefs. He has an above-average intellect, but was trained to not question certain issues of faith throughout his life. He was told to accept, on faith, certain ideas the rejection of which would cause pain to his beloved family members.

---

3. From the Internet (one of many sources for the same kind of information):
http://en.wikipedia.org/wiki/Religiosity_and_intelligence#Studies_comparing_religious_belief_and_IQ

The priest finds himself in his mid-40s one day. His atheist friend exposes him to some rigid intellectual arguments against the validity of his faith and of religion in general. He soon finds that he has no reasonable answers to any of these questions, but must instead resort to the pat responses that he has used throughout his whole life so far.

The priest begins to think about these issues over and over again and, on some level, realizes that his atheist friend's logic is impeccable.

How likely would it be for such a man – whose entire life has been dedicated to the church, and whose very identity is wrapped up in being a priest – to suddenly allow logic and reason and intellectual honesty to over-ride his loyalties and dedications and emotions, causing him to denounce the existence of a god, reject religion and drop out of the church altogether?

It's not very likely to happen.

Emotions control people's actions, and if this man hasn't found it emotionally rewarding to be rigidly intellectually honest with himself throughout his life until now, he's not going to start acting that way all of a sudden. He'll do what all otherwise intelligent but religious believers do: he'll defend his faith any way he can until he can no longer breathe.

Strange behavior . . .

Most humans – but fortunately not everyone – seem to have an inaccessible region of their brain that apparently controls the center of ritualistic instincts.

The *instinct of ritual* is not limited to us humans, but ours are seemingly endless and, when looked upon favorably, more colorful. Elephants exhibit anthropomorphic group behavior including cheering on fellow elephants while engaging in sexual activity, and they, too, bury their dead.

And then there are the worker bees and ants. But at least when they've finished with their activities, they have accomplished something useful: a beehive, or an ant farm. Humans, however, perform all kinds of peculiar rituals whose only outcome seems to be to satisfy some psychological necessity caused by group adherence to the rituals themselves. Rituals which, when viewed from a more nearly objective perspective, such as from the point of view of an enlightened intelligent creature from a distant galaxy for example, would most reasonably be interpreted as evidence of either a severe lack of education, a sign of low intellectual ability, and/or some form of mental disorder.

The following is not meant to offend anyone, but rather to look critically at our own peculiar behaviors and see how they look when we view them from a rational, more nearly objective, perspective.

Some examples of what a logical humanoid from some other planet might find to be pointless human activities include: watching, and to some extent participating in, competitive sports (such as football, baseball, basketball, volley ball, golf ball, medicine ball, cricket ball, polo ball, Ping-Pong ball, bowling ball, hockey ball, badminton ball, rugby ball, tennis ball); parades; dancing; wearing ties; going to bars; piercing things like: ears, noses, eyebrows, nipples, lips, tongues, genitalia, etc.; painting one's fingernails and/or toenails; wearing make-up; wearing currently stylish clothes while getting rid of previously stylish ones; wearing currently stylish hair; joining and participating in fraternities and sororities; celebrating national holidays (including New Year's Eve); tying yellow ribbons around trees; wearing symbolic wrist bands, wearing symbolic arm bands (including that funny-looking one with the little twisted cross that the Germans used to be so fond of); toasting "special occasions" with champagne; and, of course, the many peculiar ceremonies associated with the multitude of various religions of the world.

Hooked on mindless activities:

Some of these religious rituals include, among other things: marriages; funerals (much like our elephant friends); the bending of thy knees; praying; lighting candles; sprinkling "holy" water; confessing; making the sign of the cross; saying the rosary; refraining from eating certain foods all the time and/or certain other foods only during special occasions; wearing a skull cap; wearing a tallith; kissing a Torah (which is so holy and sacred that the face of it must not be touched by human hands); facing Mecca or Jerusalem when praying (which becomes problematic if such people should become astronauts orbiting the earth); doing a rain dance; observing the Sabbath on the debatable Sabbath day itself (which would become problematic were such people to find themselves standing on either the North or South Pole); flogging oneself; performing war dances and rain dances; allowing someone to put ashes, in the form of a cross, on one's forehead; the ritual of dunking a baby in water; sticking pins into voodoo dolls; using magic potions, amulets, tokens and fetishes; using peyote; painting faces and bodies; sacrificing animals; and, of course, sacrificing humans.

Do we have reason to believe ourselves to be the most intelligent species of life on this planet – let alone in the entire galaxy, let alone throughout the entire universe, and therefore this 'god' is in our image?[4]

Egocentrism, ethnocentrism, and nationalism —what Einstein called "the measles of mankind" – these are all illnesses of society.

Is there *anything* legitimate behind *any* of this stuff?

It seems appropriate first to examine the notion of a god— which is at the heart of the three major religions of the world— and see whether or not such a concept makes any sense.

If an atheist is defined as being a person who denies the existence of (a) god, then we are all atheists. For there are enough gods to go around that, as a group, they contradict each other. You can't be a monotheist (believer in only one god) and a polytheist (believer in many gods) at the same time. So, with respect to each and every one of the gods of which you deny the existence, you are an atheist.

The currently trendy *god du jour* is, of course, alleged to be "the 'creator' of the universe" and is the one and only such thing according to Christianity,

---

4. I know, some of you are thinking it is we who are in "god's" image. But if A is in the image of B, then B is also in the image of A. For example, there can be no such thing as a person who isn't also in the image of his own photograph. Still, the point here is that humans are so arrogant as to claim they are in the image of an entity they claim is the "supreme" being. This also assumes that evolution has somehow magically stopped with humans on this planet. Once more, all of the evidence is to the contrary.

Judaism and Islam (although they disagree with each other regarding specific religious issues— apparently just enough to justify various holy wars in the name of this all loving but vengeful god).

By this particular definition, that is, "god as creator of the universe," *I* am certainly an atheist. In fact, I'm an atheist under any conventional definition of a god.

Yet for some peculiar reason(s), some people are so anxious to believe in a god that they'll change the definition until it becomes just another word for the universe itself (which, of course, clearly *does* exist), and they'll call *that* "god" (as, for example, the religion of *pantheism*).

To the best of my understanding, at that point it becomes meaningless. For if there's no difference between the concept of a god, and that of the universe itself, then there's no distinction between the two. If there were no distinction between the two, there would be no point in using the word *god*, since the word *universe* is already there — and happily free of any unfortunate mystical, or supernatural, connotations.

[Of course, the idea of the universe having a united, yet somehow at the same time fragmented, conscious awareness that in some way connects us all (or connects all of our conscious awareness) might be considered as some kind of a god, but even by that definition, it would be more of a speculation about the nature of the universe as a whole than anything resembling any standard connotation of the concept of a god.]

So for our purposes here, I'll refer to the "one god" of the three major religions (Judaism, Islam and Christianity) when considering whether or not there is such a thing. Similar arguments, if not the same ones, can always be used to discredit any remaining gods, or god-like objects, from any of the other religions.

The chapter including the essay *How Did It All Begin?* explains the fact that the universe has always been here and could therefore not have been "created." The refutation of the existence of a god as the "creator of the universe" is already incorporated within that essay. So let's examine some other ideas regarding why one should not believe in a god.

Characteristic of the current conception of a god is the notion that it created everything. Also characteristic of this god is: its alleged omniscience (that is, the quality of being "all-knowing"), omnipotence (all-powerfulness), and "omni-goodness" (the idea that it is all-good). [Once again, an unproved and untested assumption is made; in this case, that there is such a thing as "good," that we all know what that means, and that we all agree on this meaning. To make things easier for the purpose of this essay, we'll pretend that this assumption is true (at least for now).]

Of course, if this god really existed, then it would have to be part of *everything*, for whatever really exists *has to* be part of the all-inclusive, totally-encompassing, excluding nothing, thing which we call "everything."

It would then naturally follow that if this god created everything, and is part of everything, then it would have to have created *itself*—which is, of course, a major logical contradiction. And why would it need to do that if it already were in existence anyway? On the other hand, how can anything not already in existence create itself (when it's not there to begin with)? In fact, what way can something that doesn't exist *do anything at all?*

Something that was already in existence, would have no need, let alone ability, to create *itself*. It would already have been there! So clearly there cannot be such a thing as a god that created *everything*.

Okay then, what about a god that created everything *else?* That is, a god which did not create itself (presumably it had been around forever – where? I don't know), but somehow created everything else. Well, the only *everything else* that there is, is the entire universe itself (minus this god, of course)! Until now, the universe has been understood to contain everything. But under this notion, it must be viewed as being everything else *other* than a god – which is somehow not part of the universe.

Naturally, a god who created everything, other than itself, would still have to account for its own existence. Religion always simply takes the truth about the universe itself (i.e., it is the universe which has always existed), and imposes it onto this imaginary god. Why is it *not* okay to claim that the universe – which we clearly know *does* exist – has always existed, but it *is* okay to say that this god (which is evidently permanently lost, as it can never seem to be located) "created" the universe, and has, somehow, always existed?

However, it is *the actual universe*, and not an imaginary god, which has always been in existence. Again, that aspect of this issue is already covered in *How Did It All Begin?*

So let's go on to examine some of the other alleged attributes of this god.

Most, if not every, self-professed religious person believes that there is evil in the world, as do most non-religious people, although they may disagree on some of the particulars.

If an all-knowing (omniscient) god had created everything (other than itself), and if there is evil in the world, then god created evil. There is no getting around this. That is a logical syllogism whose conclusion must be true if the premise is true.

How could any god be all-good and yet have created evil? For that matter, how could an all-good god, who is also all-powerful, even allow evil to *exist*, under any circumstances, no matter who or what created the evil? Even if, presumably, this god were no longer the creator of everything (other than itself), it would not allow anything *else* to create evil, if it were truly all good. Otherwise it's either not all good, or else it's not all-powerful.

Take your pick.

[These obvious contradictions go unanswered by religious leaders (with the exception of the previously discredited decree of "free-will," which still wouldn't explain how an all-good and all-powerful god would allow anyone or anything to bring evil into the world).]

It is difficult to believe that the Bible can be taken to be an authority about *anything*, much less about *everything*. This silly book has been around for only a tiny fraction of the existence of human life on earth, and appears, among

90

other things, to be totally oblivious to at least two of the world's most populous countries: India and China. Evidently they didn't exist. It also pretends to be a book of non-fiction and an authority about human behavior, ethics, philosophy, psychology, science, etc.

We know that Christianity has had to resort to murder in order to maintain its followers on more than one occasion.

The Salem Witch-hunts, the Inquisition, the Crusades and the persecution of pagans, Jews, and atheists throughout its history, is evidence of an inherently bigoted, totalitarian, cruel and irrational institution. Is it just coincidence that a young Adolf Hitler was a Catholic choirboy, in whose Austrian church was proudly displayed the sign of the cross—in the form of a swastika? Could anti-Semitism have flourished without religious intolerance inherent in competing religions with their individual, specific styles of dictatorial irrationality?

Naturally, the apologists will come up with all kinds of excuses for these facts of history. "Oh, that was a long time ago," some will say. Does that mean, magically, that these things never happened? Or does that mean that the current form of new and improved Christianity is totally unconnected to its ancestor?

Others will point out that many noble things have been done in the name of Christianity for charity. Yet major corporations also do charitable deeds, and both for the same purpose: publicity and good-will advertising. How often do religious-based charities perform such deeds anonymously?

Also, charity is understood to be a separate thing from religion.

That's one of the reasons why they are two different words.

One is a selfless act of helping those less fortunate. The other is a rigid, dogmatic series of codes of behavior, authoritarian proclamations regarding morality, abundantly inaccurate, vacuous "explanations" about natural phenomena, and empty, silly, anti-intellectual and conformist rituals which dehumanize its followers.

Naturally, credit is to be given to any institution that encourages charitable and noble thoughts and actions, to the extent that it does. But it only confuses the issue to pretend that charity is an automatic part of religion.

Still, others may claim that those who partook in the Crusades, the Witch hunts, the Inquisition, the lynching of black people in the Southern part of the United States, etc., were not *true* Christians—even though many of the Churches during those times were fully behind such activities (as they are still today – witness the so-called Christian Identity Movement, and the Christian Republicans – all of whom are dedicated to overthrowing the United States government and replacing it with a theocracy).

These are frequently the same people who like to use the word "christian" in place of the adjectives "decent," "honorable," "kind," etc. Perhaps they believe that changing the meaning of words will somehow change the history of the institutions and traditions behind them, and the people to whom those words refer.

In any case, it doesn't matter what they say in this regard. For if one calls himself a Christian and kills (or does something good and noble, for that matter) in the name of his religion, and/or in the name of Christ, then that is how a Christian acts. I am willing to accept the word of the individual regarding which theological doctrines they claim to believe. And I don't believe that anyone else has the power to remove from any individual whichever religious label he places upon himself.

But all of this is irrelevant to the very foundation of at least the three currently popular religions: namely, the existence of a god.

Can any reasonably intelligent, reasonably well-educated, mentally healthy person *honestly* think that out of the billions of forms of life, ranging from the mighty elephant to the microscopic bacterium, currently existing on this tiny planet—which is one of billions of planets which likely contain life in this galaxy, which is one of billions upon billions of galaxies throughout the universe which also likely contain life—that this one, currently popular, god/creator of the universe somehow would be in *our* petty little image?

Could this be reasonably interpreted as being perhaps a bit pathetic, or arrogant, or perhaps absurdly egomaniacal? There's a god in our image? A god who created most of the living beings on this planet way before humans ever evolved was somehow in our image?

And *why* do people believe such a thing? They do so because their mother told them so, and they want to believe it. It is this fact, and the fear of death, that reinforce these mutually exclusive and unjustified belief systems.

After all, what is the difference between blind religious faith and wishful thinking?

Ultimately, nothing whatsoever.

It is the very concept of blind faith upon which religion rests, and for which it rewards its followers. The idea is that this god created everything and gave people "free will" to believe in it, against all evidence, logic and reason to the contrary.

Religion is an emotional issue. It clearly cannot stand up as an intellectual one.

god is the Santa Claus of adults.

Adults react the same way, when confronted by evidence disproving the existence of a god, as children do when they discover, for the first time, that their parents lied to them about Santa Claus. In both cases, simple, but untrue, stories are invented to explain some things. In both cases, the people who are told these stories want to believe them to be true, and they allow their emotions to over-ride their intellects and make unreasonable assumptions about reality that they know would violate physical laws.

This phenomenon is understandable in children.

Yet there have always been simple stories to "explain" events to adults that are far more complex than the stories that "explain" them. There is nothing new about that. Each religion is seen as being "the one true religion," but only seen that way by those who follow it, and not by those who don't. One man's cult is another man's religion—the status of which has always been determined by its individual popularity and by nothing else.

How did we get into this mess?

If we were merely average students of history, we'd be aware of the fact that monotheism (the belief in one big, magically all-powerful god) grew directly out of polytheism (a belief in many little ones). The evolution of religion may have started out honestly, and may have been relatively innocent (although I personally doubt it). In any case, it didn't take long for these simple "explanations" to be used by priest-politicians for the purpose of controlling the masses.

They used threats of some mystical after-life retribution for disobeying their god's earthly commandments. The various Kings, Emperors, Tzars, Popes, etc., each claimed to be their god's representative on earth, in order to bully people into accepting their right to govern the masses. Naturally, *some* of the laws imposed by these rulers were reasonable (not killing, not lying, not stealing, for example).

Yet, does a society, or culture, act fairly or responsibly when it coerces its populace into conforming based on intimidation, fear, lies and threats?

Why hold the religion industry above rational criticism? Why not call a spade a spade? Is it reasonable to continue supporting any such institution whose very foundation is based on deceit, oppression and murder? Is that not the case with religion?

How have modern religions evolved?

Once upon a time, there were the gods of lightning, thunder, love, war, etc. Eventually, many little gods were condensed into one. Then it became a sin, under Judaism's ten commandments – which includes not coveting thy neighbor's oxen (nor his ass, for that matter) – to have any false idols (as if they weren't all false). That paved the way for monotheism as a political device to manipulate and control the masses.

It is no small matter – no pun intended – that both Judaism and Islam command their male followers to be circumcised. If one is willing to cut off the tip of his son's penis, when his son is far too young to give his consent, then that person's religion has left an indelible mark on the offspring who becomes yet another of its victims – a living symbol, unwillingly branded for life as part of the clan.

How is that *not* child abuse?

In spite of things like this, religion often proclaims itself to be the ultimate moral authority – especially in all areas related to sex. But they are, in reality, the ultimate moral hypocrites. After all, if you will allow them to tell you whom to have sex with, how often, what kind, under what circumstances and for what purpose, then they really do have you by the genitals!

The *Ten Commandments* were allegedly given to Moses by his god. The reason people might have found a Moses to be so holy and sacred, was that he supposedly heard the "voice of god" coming from a burning bush. Of course, in those days, the science of psychology was non-existent. People were unaware of the mental disease known today as schizophrenia, one of whose main characteristics is having hallucinations (for example, hearing voices inside one's head which are not attributable to any outside source). So to the people hanging around Moses, this business about hearing god's voice coming from some flaming shrubbery would have seemed nothing short of a miracle. They wouldn't have known that such a person was a very sick man.

Perhaps the biggest mistake of the Jewish religion was creating the idea of a messiah.

That device allowed the early Christians to claim that Jesus had already been predicted by the Bible to be the savior of the human race. Because of that, the Christians were able to usurp the political and psychological power of the Jews as the rulers of the masses in their own little corner of the world.

The Christians could simply claim something like: "Hey, Jews! You guys predicted the coming of the messiah, yet you missed him. Now you'll go to hell unless you follow *our* one true religion which recognizes, decades after his alleged death, that Jesus was the savior."

Curiously, in the sequel to the Bible, the New Testament, the personality of the god described therein completely contradicts that of the original book. And the book ends by warning its followers to ignore any *other* books that may come later and have anything *else* to say about these religious issues. Of course, the Church of Latter Day Saints of Jesus Christ, a.k.a. the Mormons, has done just that. They have *The Book of Mormon*, which came after the New Testament, in which it is alleged that Jesus Christ actually discovered America before the Native Americans, the Vikings and Columbus.

They also believe that the sacred writings of their god came to them in the form of plates that fell to earth from the sky, although nobody seems to know whatever happened to those plates. [Rumor has it, though, that after these plates were "translated," they mystically evaporated unto heaven.]

Then there is the problem of "creation."

"Original sin" (which seems to me to be anything but original), into which we are all supposedly born, refers to Eve, "the first woman," eating from the tree of knowledge (of good and evil). The Sistine Chapel even commemorates this myth on its ceiling in the Vatican in Rome.

This is one of the basics of most modern so-called "Western" religions – especially Catholicism. "Original sin" blames Adam and Eve for eating from the tree of knowledge of good and evil, which is yet another contradiction. Before they had eaten from that tree, they would have been innocent of knowing the difference between good and evil and so wouldn't have been responsible for making an evil decision or taking an evil action (eating from that tree against god's commandment). And why wouldn't god have wanted them to know the difference between good and evil anyway? Isn't this character supposed to support all that's good?

And this god, in its 'infinite' wisdom couldn't foresee what Adam and Eve would necessarily have to do, even though this god were all knowing?

So much for original sin.

But something very strange happened towards the end of 1996.

Pope John-Paul II publicly admitted that *evolution is a fact*. Of course, it is; but doesn't that just blow the whole idea of "creation" to hell? Does anyone detect some logically irreconcilable inconsistency here? Aren't we supposed to be "sinners" based on the Adam and Eve myth? (Oh, I know the apologists will come up with all kinds of after-the-fact excuses. They always do. They'll say things like "Well of course evolution is a fact. We know that, but god made the evolution which created Adam and Eve." Or they'll say that: "Adam and Eve were not to be taken literally – that much of the bible is just figurative. Of course, the material about the existence of a god isn't figurative, and Moses turning a stick into a snake, getting the Ten Commandments, and parting the Red Sea isn't figurative. And Jesus turning water into wine and bread into fish and being born from an 'immaculate conception' [sic] isn't figurative." Until one day when each of these nonsensical claims will be buried once and for all alongside of the currently unpopular gods of history about whom hardly anyone still gives a damn.)

What about other odd religious proclamations? Christians claim that their god is really a man, the son of god, all human, yet not human at all, who was called Jesus, and who allegedly *"died for our sins."*

But just try to get them to tell you exactly what any of that means. What *does* that mean? Which sins? How can a god die? Aren't they supposed to be immortal? If a god *could* die, then it would still be dead today; after all, you can't be both dead and not dead at the same time – they *are* opposites! In what way can anything die *for* our sins? How could Jesus have "died for our sins" when we hadn't even been born yet? If a god died for our sins, wouldn't it then be incumbent upon us to commit some? If not, what precisely *did* it die for then? Which sins have we committed? Is birth itself a sin? How can we be 'born into sin'? Is it by mere proclamation? If that's so, then I proclaim just the opposite! Now what do we do? Does this sin business become legitimized simply because it is written in the bible? Then *this* book will serve as *my* bible, in which it is written that there is no one born into sin.

Then there are other questions which, when pondered aloud, may make the individual worshipper a little bit more uncomfortable as he begins to realize the inherent meaninglessness of his religion.

For example:

Can Roman Catholics *honestly* believe they are eating the body and blood of Christ – for real – during their holy communion? Why would they *want* to cannibalize their own god anyway? Why would Episcopalians follow the same strange ritual, but at least acknowledge that it's just symbolic and they're not *really* eating their god?

Everthemore, all truly believing Christians still have a lot of explaining to do, at least to themselves, regarding the highly bizarre claims they make about their human god, Jesus Christ.

How could Jesus be his own son?

How could he be his own father?

What does that mean?

How could there be more than one god at the same time when there is only one god? What exactly *do* these claims *mean?* If Jesus were "the son of [a] god," then why call him a "god"? And if he "died for our sins," then he is dead! Oh no, they say, he came back to life. Really? Jesus is a zombie? Then he didn't die, did he?

He couldn't have died "for our sins," whatever that means, unless he died. That's certainly clear. And if he isn't dead, then he didn't die – (that's what it means: to die is to be dead)!

Furthermore, if he were alive, then he'd have to be living, wouldn't he? So he wouldn't have died at all, and therefore wouldn't have "died for our sins" (whatever that means), or for anything else in that case. If he were living, where would he be? What would be his address?

If the typically evasive reply is, "it's a mystery," then the response is: "The real mystery is how anyone can honestly claim to believe such vague, empty, meaningless nonsense."

Another very peculiar, yet typical, reply is: "There are just some things that we, as humans can't know . . . and here is what they are . . ."

One can't talk about what it is that can't be talked about.

Similarly, one can't know what it is that can't be known. Furthermore, one cannot know what kinds of information, if any, cannot be known. Even if it were true that certain kinds of things couldn't be known, you could never *know* exactly what it was that could not be known (since it could not be known). All you would be able to say about that, with near certainty, is that *you* don't know some specific things at this point in time, and perhaps you never will. But how could you possibly make a claim about what absolutely cannot be known – especially by someone else (who does know it)? How can anyone honestly believe that they have the right, let alone the ability, to penetrate the mind of another person and accurately account for the contents of that person's head?

How could you *know*, with certainty, what you're claiming can't be known with certainty? But most of all, how can anyone be so pompous and arrogant as to tell you what fits into the list of things that we, as humans, just can't know – while at the same time claiming that he, who is also presumably human, *does know* what it is that fits into this list of things that we can't know?

And then there are some (Pentecostal/Apostolic) Christians who even "speak in tongues," which sounds more promising than it turns out to be.

During these tongue-speaking times, the participants, in a religious fervor, utter a stream of sounds emanating from their mouths, which do not convey any actual meaning and therefore can't be translated into real words in any known language.

But Christians aren't the only victims of religion.

Orthodox Jews, for example, are not allowed to "work" on the Sabbath day that, in their religion, is from sundown Friday night to sundown Saturday night. They may walk twelve miles to a synagogue, where they'll spend the entire day in prayer to a god who presumably already knows what's on their minds. They may walk down ten flights of stairs, but may use neither an elevator nor an escalator. These activities are not considered to be work. And orthodox Jews can't *drive* to synagogue, because that uses an electric engine that is considered, somehow, to be some sort of version of "fire," which, naturally, is not allowed, either.

Because of their interpretation of "work," they cannot turn lights on or off and are not allowed to cook on their Sabbath.

And during the special High Holy Days, the orthodox Jewish believer knows that his god is writing down his name in some supernatural book of life, while the reformed Jew acknowledges this to be merely metaphoric. On the Day of Atonement, they may neither eat nor drink for a twenty-four hour period (during which time, it would be a sin). Then, suddenly, it's okay to eat and drink again (providing you're not eating something prohibited by the religion)!

And on the festival of Passover (during which Jesus is said to have had his last supper), the Jews celebrate by using a ritualistic reminder that their ancestors were forced to wander around aimlessly in the Egyptian desert for forty days and nights with really lousy food including "unleavened bread," or matzo (which is the same material out of which are manufactured the holy communion wafers).

So to commemorate how awful it all was, every year during Passover they eat the same kind of unpleasant food for dinner, including the mandatory symbolic bitter herbs, frequently after having subjected themselves to an almost unbearably lengthy series of prayers.

To the severely orthodox Jew and Muslim, women must be separated from the men in their house of prayers, and women are always subservient. (This is also true of the many of the extremist Christians.) And the orthodox Jewish male may not trim his sideburns.

Of course, neither an orthodox Jew nor Muslim may eat pork, nor shellfish, nor combine diary products with meat on any occasion.

Wait a minute!

Before anyone inaccurately accuses me of being too harsh, let me interject that I have absolutely nothing against individual religious people—provided that they are otherwise decent, good, fair and sympathetic. For it's the sin (of religion) that I dislike, and not the sinner.

[The exception is, of course, when the sin *is* the sinner—such as those self-righteous right-wing religious conservatives (whether Christian, Jew or Muslim) who try to impose "their" views on others—especially through laws.

They are the ones who try to prevent funding of the arts, are against the fact that abortion is legal where it is legal, try to ban books, records and movies and attempt to turn off *your* television set when something comes on which they don't want you to see (usually it's about sex, because many of these people wouldn't recognize a good orgasm if it blew their legs off)!]

There are many questions to be asked which are more reasonable than the answers you're likely to get when it comes to the subject of religion.

For example, what is meant by the use of the word "soul?" Where would such a thing be located? What would it made of? Exactly what is it that you "believe" in when you claim to believe in the "existence" of a soul? If you can't answer these questions specifically, logically, unambiguously and accurately, then exactly what *do* these concepts mean to you?

Could it be that you simply have a *feeling* that you actually believe in something that has real meaning, whereas in reality, your apparent "belief" isn't simply false, but is actually meaningless?

How could there be any kind of superior god when evolution progresses from the lesser to the greater, and such a being wouldn't have evolved yet? Why would an allegedly superior god have the psychological characteristics of a mentally disturbed child, always needing to be re-assured and praised, acknowledged and worshipped? Why pray to an all-knowing god? Wouldn't this obvious redundancy just be annoying to it?

Where would this god be located and what would it be made of?

Is this god a male? In what way could this god be a "he" unless it had a penis? What would a god do with a penis? (Stick it in his ear? After all, a god wouldn't need to urinate, neither could he have sexual intercourse unless there were a goddess or he went outside of his species, engaging in what would amount to celestial bestiality. Come to think of it, if Jesus were the 'son of god,' then that would mean that there was some supernatural bestiality engaged in by the same god who gave us our hysterically Puritanical sexually frigid 'morality.' Better still, why would a god need to go through the regular channels to have a 'son'? Why not just create one in a fraction of a second, after all, this all-powerful being made the world in just six days – after which it was so all-tired that it had to rest – so why wouldn't it be able to just pop a Jesus into existence in a fraction of a second? That would certainly have been more impressive and 'miraculous' than forcing himself on an unsuspecting married woman.)

How could there be a heaven?
Where would it be located?

How would anyone get there – especially once they're dead, buried in a wooden box, submerged underground, covered in dirt, and busy rotting? Or are we supposed to ignore all of this overwhelmingly obvious evidence to the contrary, and pretend that this dead person isn't really there at all?

If there were a god who wanted his creation to believe in him, then why wouldn't he only have created believers?

Why would an egotistical god, who wanted everyone to believe he existed, provide only evidence to the contrary – proving just the opposite? How is it that the Catholic Church gets away with referring to the office of the Pope as being the "Holy See?" [He can see no better than I can, in any respect, metaphoric as well as literal. And nothing is "holy."] Try answering those questions.

I know that people get very defensive about "their" beliefs. I, personally, didn't cause there to be no god. In fact, I'd be willing to vote for the existence of such an entity, were it possible. I'm just not willing to pretend that the Emperor is wearing any clothes, when, in this case, there isn't even any Emperor.

And then there is the idea that one cannot criticize someone else's religion.

Why not? Could it be because those within the religion are *already* within the religion; therefore how severe, or objective, would *their* criticisms ever be?

In fact, isn't it an inherent conflict of interest to judge one's own religion from within the religion itself?

One absolutely *can* criticize someone else's religion. I am an atheist, and as such would have to remain silent on all of these issues (as many people might wish, anyway) if I followed the rule about not criticizing someone else's religion. [Furthermore, it's perfectly fair and reasonable to analyze critically what, for too many centuries, has had a major and negative influence upon us all. It has been a blight on society by crippling creativity, logic, reason, and progressive, independent thinking. It has also imprisoned the ultimate happiness of the masses – which have been, and continue to be, perhaps not entirely surprisingly, the ironic victims and biggest supporters of religion.]

What about "the argument from design" as an excuse for the existence of a god?

This is the tired notion that, for example, if we were to find a watch on the ground, we wouldn't believe it had just evolved—we'd know that there was a watchmaker who had consciously designed it. Similarly, as the theory goes, since the structure and function of the human being is so obviously complex, and fits very well into his environment, then he, too, must have been "designed." Therefore, there must be a "designer" (that is, a god).

There are some logical problems with this argument.

For one thing, we already know in advance that watchmakers design watches, so that would be stacking the deck unfairly. A better example to use would have been a rock. The rock was not "designed," obviously, by anybody. But even in the case of the watch, it is perfectly reasonable to talk of it in terms of evolution. For *everything* – including we humans and everything that we cause to happen – is part of the on-going process of evolution, or change.

Watches, like everything else, have evolved. But, more importantly, so have we. The evidence is irrefutable regarding the evolution of all forms of life—including humans. We have fossil evidence (much older than the age of the earth according to the Bible) connecting us *Homo sapiens* to our ever-evolving ancestors—who no longer reproduce their species—thereby proving the theory of evolution by having slowly turned into modern day humans.

Another way of looking at this issue would be to say that if you need to believe that we were "designed," then think of us as having been designed by the unconscious process called evolution, rather than by anything with purposeful intent.

Within our genetic material is a trigger mechanism, which can cause random mutations that are then passed onto our offspring. If these mutations were so negative that they would prevent us from effectively existing within our environment, then we'd die—after which, most of us would have some difficulties reproducing.

However, if these mutations were useful in allowing us (or any other animal or plant) to adapt to our environment, then we'd be alive and able to reproduce offspring who would have the same genetic characteristics. This is a concept explained by Charles Darwin.

Then there are neutral mutations that do nothing other than increase variety. Changes that produce negative results tend not to get reproduced. Changes that lead to positive adaptations and improvements get passed on. Eventually, when enough genetic changes have occurred, a new species has evolved. It isn't a fine line that separates one designated species from another; it's a gradual and blurred line.

Greatest Conceivable Being?

Then there's the ontological argument for the existence of a god. This is the peculiar claim that a god exists because it is defined as being "the greatest conceivable being." And, after all, if one can imagine "the greatest conceivable being," then such a being must exist; because a "real, *existing* greatest being" is greater than a similar being that does *not* exist. [From this, it follows that a non-existent god, *in reality* is more non-existent than the *conception* of a non-existent god.]

One can easily shoot down this theory. It's the same thing as saying: "I wish to have a million real dollars, in reality. Of course, imagining to have a million dollars is not as real as really having a million dollars; therefore, I really *do* have a million dollars."

In each example, we beg the essential issue. Of course a "real god," who really *did* exist, would be greater than an imaginary one that really *didn't*.

And my really having a million dollars is more real than my imagining to have that kind of money. But neither concept becomes true simply by defining it into existence! It is not *we*, after all, who are supposed to be all-powerful.

Is god love? Is god everywhere?

Then there is the empty slogan, "god is love."

I don't believe that this definition is strictly out of the '60s, although it seems very transcendental and flower-power-like (which is not automatically a negative). But the idea that "god is love," is, once more, absolutely meaningless! What does it mean? Love is love. Love is an emotion. Is god also an emotion? Or, rather, do people now admit that 'god' is simply an emotional issue rather than a real thing? What exactly is this 'god' then? Is it something sticky?

What, if anything, does "god is love" mean?

And, while we're at it, what does it mean to claim, "god is everywhere?" Certainly if one thinks that 'god' were the 'creator' of the universe, then it couldn't possibly be everywhere! For that would mean, yet again, that this god created itself.

But the problem remains: how could a real god really exist unless it were part of everything (the universe)? The only thing that is not part of everything is nothing.[5] Yet if it were part of everything, then it certainly could not have created everything or it would have to have created itself, as we've seen before -- which, of course, isn't a rational concept.

---

5. Of course, 'god' really is nothing.

So, again, what meaning is there in claiming that 'god' is everywhere and in every thing? That's an odd statement that avoids the questions: What *is* this god that is supposed to be everywhere? *What else*, aside from the universe as a whole, could possibly be located everywhere – especially considering that "everywhere" includes "all of time"?

Those who make such a statement that god is the universe, are not Christians, Jews, Muslims, etc. They are Pantheists – and Pantheism has no clear meaning because there is no clear distinction between the definition of god and that of the universe itself.

And then there are these other unanswered (perhaps unanswerable) religious questions:

If there *were* such a thing as life after death, then how could there be *death?* What does the phrase "life after death" mean?

If it means, and this is the only rational meaning I can come up with, "the continuation of an individual's conscious awareness after that person's body dies," then how can it possibly continue to survive upon the termination of brain functions?

Conscious awareness is merely the combination of our sensory impressions, memories and thought processes. All of these things require a functioning brain and central nervous system in order to exist in us humans in the first place. Break someone's spinal cord at the neck, and if it doesn't kill him, he'll be paralyzed—and unable to feel from the neck down. That person is no longer consciously aware of his body below his head (except for his ability to see it and think about it), and he's not even dead yet!

Similarly, someone who has had a stroke (that is, has had a part of his brain actually die) may lose his sense of sight, or of hearing, or have other losses of brain function, causing damage to his conscious awareness. Slice off someone's brain, one layer at a time, and you will destroy that person's conscious awareness, bit by bit, until it has been eliminated completely—and this will have been done during his lifetime.

Whoever doubts that our conscious awareness is a purely physical (electrochemical) process, must never have learned about the effects of drugs (such as alcohol, LSD, marijuana, cocaine, etc.) on perception. Perhaps these people are unaware of the effects of drugs (such as sleeping pills, caffeine, etc.) on states of awareness. Maybe they've never heard of some of the effects of having a brain tumor. Perhaps they're unaware of how certain brain surgery causes changes in perception, awareness, memory and behavior. Perhaps they've forgotten the affects of strokes, which kill off brain cells and affect memory, sight, thoughts, and other areas of conscious awareness. Maybe they've forgotten that when we sleep, but are not dreaming, we are not consciously aware of anything. [Of course there is a time during sleep when we are aware consciously even though we are not dreaming, but there is also a time when we are not.]

Since one can lose some, or even all, of one's conscious awareness *during one's lifetime,* how then would it be possible – even in theory – for an individual's conscious awareness to survive one's own *death?* How could conscious awareness continue after someone's head has exploded, for instance? How is the idea of surviving one's physical death anything *other* than meaningless wishful thinking? Of course most of us would like to be immortal, but, once more, reality is not determined by a democratic vote.

When it comes to religious issues, as with all other issues, give yourself permission to think. Any real god and/or true idea should easily and proudly be able to stand up to tough scrutiny – if it can't, why clutter your brain with it? Why give it any power at all?

Here is an imaginary dialogue between a religious parent and an inquisitive child with a functioning brain:

"If god made everything, and if he knows everything, then he must know how the people he created are going to act before they do anything. Right Dad?"

"Right, son."

"Then god should be blamed when we, his 'creations,' do evil things."

"No son, that's why he gave us all free will."

"What does that mean?"

"Well, son, it means that he gave each of us the ability to choose to do good or evil."

"Then if an all-knowing, all-powerful god made us, and if we do evil, then he knew he was creating something which would do evil. So he made evil."

"No! He didn't make evil!"

"Then god didn't make *everything*, did he? Either that, or this god is not all-knowing."

"Yes he created everything and is all knowing and did not make evil."

"But that doesn't make any sense."

"You don't understand!"

"I don't understand."

"You're too young."

"Too young for what?"

"Too young to understand these things."

"That's not true – it's not because I am young that I don't understand; it's because it doesn't make any sense."

"That's the beauty of it all. It's a mystery. Now stop asking me these questions. I have a headache. Go to sleep."

"Good night."

"But god isn't religion, and there is a difference between religion and faith."

For some strange reason, there are people who will agree that "religion" is dangerous, but will claim that what *they* believe is neither a religion, nor dangerous. Then they proceed to define some 'god' in which they believe. This 'god,' of course, exactly fits the very definition of a religion. However, they will simply claim it to be a "faith," ignoring all attempts to explain that faith is a synonym for religion.

Others may deny their religious belief by calling them a "relationship" (almost always to some magical god or a son of a god, or something along those lines). Yet this is always a "relationship" to something unreal and, naturally, always of a religious nature.

People, they admit, created religion. That is why theirs (usually some form of Christianity) is somehow not a religion, but a "faith" in the "real thing." The truth is that all gods throughout history, including all the current ones in existence (as concepts), have always been religion. That is what religion is. That is what it has always been.

This currently popular trend of simply imposing the title "not a religion" onto the series of beliefs which clearly constitute religion, by definition, is just more evidence of the dishonesty inherent in the very nature of beliefs of this kind. Frequently religious people intentionally bastardize the language by purposefully using words in sloppy ways and trying to shift distinct meanings into fuzzy ones so that they can make certain untrue claims appear to be less obviously insane.

Then there is the problem of faith itself:

Consider the matter of *blind faith*. Not faith based on observation, logic, reason and evidence, but faith based on the hope that what other people claim to be true actually is true. If it were true, then it wouldn't fall apart when held up to the light of reason. Yet this is precisely what happens in the case of the metaphysical and supernatural claims of religion.

On the surface of it, and with no further examination, there appears to be nothing wrong with faith. After all, before any problem can be solved, we must start with the assumption that it can be solved. This is a kind of faith. It's the type of faith that is reasonable, because if we assume the opposite (that is, if we have *faith* that it *cannot* be solved), then it definitely *will not* be solved.

Of course, *assuming* it can be solved doesn't guarantee that it *will* be. But assuming that it can't be, guarantees that it won't. That means that a certain type of faith is a *necessary, but not sufficient,* condition for, among other things, solving scientific problems and accomplishing one's goals.

There are two different kinds of faith.

There is faith based on observation, logic, evidence and reason (such as faith in the consistency of the laws of nature, faith in science and, to a lesser degree, faith in individual people whom we know, based upon our past experiences with them). This kind of faith is reasonable as it has its basis in observed facts in the real world.

And then there's religious or blind faith (such as faith that you'll win the lottery, faith in the existence of a god and in one's religion, faith that everything will happen exactly as we want it to).

This kind of faith is not reasonable.

There is no difference between religious *(blind)* faith and wishful thinking.

And when used as a device to replace human action, where such action could be helpful, it becomes a highly destructive force.

This is seen, for example, when people calling themselves "Christian 'Scientists'" (as opposed to Jewish lawyers?) refuse to allow their sick or injured children to be treated in a hospital. Instead, they rely upon blind religious faith as they pray to their god while watching their child die. [Their religious freedom is, under such circumstances, imposed upon innocent children who are not only deprived of *their* religious freedom, but, ultimately, of their very lives.]

Is this kind of faith something valuable?

If it were, why wouldn't people use it in real life, when doing every-day practical things such as buying automobiles and houses? Would you drive your car totally blindfolded, based on faith? Would you send signed blank checks through the mail because you had faith that it would get there and that the recipient would be honest? In a large metropolis, would you live in a house with no door locks because of your faith in the honesty and integrity of your fellow human?

No. Because you know that kind of faith is not a rational thing.

How is it, then, that this cheap device called (blind) *faith,* which we know isn't good enough to use on less important things, *is* used by some people upon which to base their entire lives? Presumably, this should be a little more important to them than an automobile.

"Oh," but they'll tell you, "I don't have faith in people; I put my faith in (a) 'god'!"

Well, that's what they'll say, of course; but every single god that has ever been written or spoken about has been an idea transmitted to people by people. So it is necessarily the case that no matter what the religious person claims, his faith is in *people* and not in any god, or gods -- but only in those people who have told him about whichever god he wants to believe exists.

At what cost do these believers, with the (blind) faith of a child, become members of the flock of sheep that (at least in the case of Christianity) their own religion openly calls them?

The cost is to their individuality, their honesty and integrity – and sometimes to their very lives. It is at the expense of their one and only brief universal appearance, to which they have devoted a lame theology in exchange for the artifice of "feeling comfortable" and not having to examine their own lives thoroughly. Nor do they have to think about, and account for, their actions as individuals.

The problem is that most people spend virtually no time forming (let alone examining) "their" ideas, which have been inserted into their heads by other people to begin with (that is, by their family, religion, friends and society). Yet these same people spend an entire lifetime defending those views – an incredible and self-destructive waste of energy!

Why Do They Believe?

Why do some people persist in 'their' beliefs in a magical god, etc.? Well, this isn't too difficult to understand. Those people want to believe. They don't want to be dead forever some day, so they pretend they won't be – even though, as we've seen, there's no need to do so since nobody's ever going to be aware of their own death (as it's not something that occurs during one's lifetime anyway – so you'll never be thinking to yourself, "Gee, I'm dead now. Isn't this boring?").

Conscious human beings, who are aware of anything at any time, must necessarily be living.

Some people pretend that "anything's possible" (sic) and that perhaps once they die, they won't really be dead, but will somehow go to 'heaven' – the Disneyland ® of the dead. The magical land, somewhere over the rainbow, where the great and powerful Oz can and will grant their wishes (even though all of their wishes require attributes of physical existence which supposedly isn't part of this magical land).

Additional Problems With Religion

While freedom of religion is properly considered to be a valuable and positive "right" of the individual, in the same way and for the same reasons as is freedom of thought and expression, it still remains that there are too many negative effects on society, as a whole, which are directly attributable to this multi-billion-dollar-a-year industry. If an individual wants to practice his or her own religious dictum, that is perfectly fine and no reasonable person should have any problem with that – no matter how silly a reasonable person may find this to be.

The problems come when religious dogmatic doctrines are imposed upon society at large, especially through the legal system. This is particularly noticeable in theocratic states, and to a lesser degree from time to time in the United States of America, where the far right-wing religious fanatics practically own, and clearly pull the strings of, the Republican Party.

Putting aside the violations of their tax-exempt status as a religion (which, at least in the U.S.A., should have no tax exemption), they are clearly both religious (in their viewpoints) and, what's worse, political. They have nothing but contempt for liberalism, progress, and advancements of society as well as for individual freedom. And, if they had their way completely, these religious extremists would recklessly turn the USA into a *de facto* theocracy through stealth and deception. (Which is something we can clearly see them attempting as they encourage their minions to infiltrate politics at all levels, starting with local school boards and ending with the Supreme Court).

As it's a country whose majority populace call themselves Christians, the dangers of this, and of other extremist groups, lie in the fact that most voters cannot be objective about their goals. Most Christians cannot (and will not) separate themselves from their religion long enough to look at the dangers it is causing to the political and social fabric of their society while, typically, claiming to be doing exactly the opposite.

If such Christians were to imagine a similar group of orthodox Jewish rabbis, for example, forming a Jewish Coalition for the purpose of making sure that American citizens don't have the legal right to eat pork or seafood, and imposing circumcision by law on all male infants, how long do you suppose they would sit back and tolerate such activities in the name of freedom of religion, or speech?

Yet it is precisely the same kinds of activities in terms of imposing their religious views onto everyone else (regarding abortion, sexuality, marriage, etc.), in which extreme Christian elements engage while mixing in politics and pulling the strings of their puppet political party, the Republicans.

Not content to tell their own people, who 'freely' follow, what to do, these religious dictators insist upon doing whatever we allow them to get away with politically. And it doesn't stop there.

Due to the inherent conflicts between the opposite approaches of science and religion in dealing with the realities of the modern world, there continues to be opposing and dangerous consequences of the influence of extreme religion on technology. The scientific world is on the verge of being able to clone humans, having been successful at the cloning of other animals – including other mammals.

Soon it will be possible to clone a side of beef, for example, without having to clone the entire cow itself. In theory, the beef could be developed with no cholesterol and even with daily requirements of fiber such that there would no longer be any need for anyone to be a vegetarian if they enjoyed eating meat at all. During such a time, there would be no health reasons for being a vegetarian, nor would there be any moral reasons (since no cow would be killed for their meat – only a section of the cow would be grown in a laboratory and this living organ would have no conscious awareness and no knowledge of its existence and would make eating such meat the moral equivalent to eating a vegetable).

There is also the clear possibility of developing individual human organs without having to clone the entire human being. Such possibilities could easily lead to the end of much human suffering in many areas of human health. And then there is stem cell research, which could allow the discovery of the cure for cancer, diabetes, heart disease, etc. Yet something stands in the way of this progress.

Posing as the arbiters of what is and isn't moral (and too frequently being examples of the very definitions of hypocrisy and immorality), religious shepherds around the world have come out against scientific progress – once again – in this instance as in so many others in the past. In their naïve ignorance, they have done everything they can to stop what they falsely proclaim to be "unnatural."[6]

6. Unnatural? If something occurs in nature (which all real events do), then, by definition, it is natural. The only things that can logically be considered to be unnatural are things, which go against nature. The only things that go against nature are things that do not happen – therefore every real occurrence must be considered to be a natural one, as everything (that is real) occurs in,

To the degree to which we notice such self-proclaimed moralists succeed, we will also be seeing the further continuation of, and an ever-increasing growth in, the physical and intellectual (as well as "spiritual" in the sense of the human personality) poverty of the world's masses.

Eliminating diseases requires medical research, not mindless theocratic proclamations. And there are other good reasons to encourage additional research and development in the field of genetics, as one example.

Cloning entire humans, though nowhere nearly as important as stem cell research for medical advances, nevertheless would permit people who wish to have children, but who may otherwise be unable, to raise a family. If a specific person were cloned, it would be the exact same genetic entity as an identical twin of that person, with the exception that the clone would be that much younger. [For example, if a man in his 40s were to have an infant of him cloned, the identical twin would be forty-some years younger than he.] As has happened with all such progress in the past, this will occur in reality in the not-too-distant future.

Breakthroughs in genetics have already made it possible for older women to conceive at later stages in life than was ever possible before. It has also made it possible for women to carry the embryos of other women who are unable to become pregnant. Other scientific breakthroughs have made it possible to avoid abortions by disconnecting the zygote from the uterus through the simple device of swallowing some pills. Yet again, many religions stand in the way of public distribution and education in connection with "the morning after" pill. Ironically, they'll claim to be against abortions, but every opportunity they have to take any practical steps to help prevent the need for such operations is squandered, bitterly opposed and ultimately sabotaged.

Put simply: religion is all too frequently anti-nature, anti-sex and, above all else, anti-reality.

The Pope (who is, after all, the dictator of a theocratic state which discriminates against all other religions) and other religious leaders consistently

and is therefore a part of, nature.

rail against birth control, abortion, sex education, condom distribution, homosexuality, and any sexual activity outside of marriage (and many kinds of sexual activities within marriage as well). Unchallenged ideas against sexuality are ignorantly passed on from parent to child, frequently with dreadful consequences.

Many teen suicides are committed as a direct result of the hateful ignorance religion expresses regarding natural sexual inclinations. The point to be made here is that, once again, religions' effects on society are by far more harmful than helpful and raise the following legitimate philosophical question:

Why should we respect religion?

Of course it should be made clear now, once more, that I have nothing against any individual religious person – especially one who does good in terms of helping others. But this is admirable only to the extent that such work is not done as a publicity stunt for good-will promotion/advertising of their religion. Which leads one to wonder:

Why doesn't the Pope (this one, or any of them) spend all, or at least most, of his time doing worthwhile charitable deeds? Why doesn't the Pope copy Christ in a vow of poverty and go around giving money from the church to the poor – instead of it being the other way around?

The answer is simple. Charity is *not* religion. They are two different things. Again, I support charity. I cannot support religion.

Some people assume that we must respect the religious views of others or else those people who are, themselves, intolerant, will accuse us of being narrow-minded.

This, of course, is faulty thinking.

We can attack the content of religious claims, and do so very strongly, without attacking individuals who practice their religion (as long as they do so in ways that do not harm other people). We can support the individual victims of religion and wish them only the best while sharply attacking those elements of their (or any other) religion that we find to be irrational, dishonest, harmful and vile.

116

It is not promoting hatred of a people when one specifically only critiques (even if harshly) the claims of religious institutions. We may well have sympathy for those followers whom we may also see as being victims themselves.

But in the case of anti-Semitism, for example, we see the persecution of Jewish individuals (or people that are labeled as being Jewish based on invalid racist proclamations). It is not, for instance, a critique of the content of the Jewish religion. It is, instead, purely irrational, illogical, old-fashioned racism, prejudice and bigotry.

Naturally, there is a reason for this.

Judaism is directly responsible for both Christianity and Islam. Remove the "validity" of the content of the Old Testament of the bible--that is, remove the Judaism itself-- and you remove the very foundation of these other religions. You pull their carpet out from underneath them.

So, for that reason, they never attack the content of Judaism, but rather they attack both the Jewish people themselves and also the differences between Judaism and their own religion. Ultimately, those fanatics of other religions, who have ended up attacking Jewish people themselves (and attacking those who are simply designated as being Jewish), have done so with no rational basis whatsoever for their attacks while ignoring any valid arguments that may exist against the claims that the religion of Judaism makes about the nature reality (e.g., the existence of a 'creator of the universe').

At the end of the 20th Century, in parts of France, the far right-wing, unsatisfied with the nearly total destruction of the Jews of Europe during the nazi era of World War Two, have visited cemeteries for the sole purpose of digging up the bones of dead Jews in order to "kill them again" by smashing their bones into pieces with sledge hammers.

At the end of the first decade of the 21st Century, Islamic militants in France have also murdered innocent Jews for no logical reason.

In Germany – having nearly completely run out of Jews to kill – the extreme right-wing neo-nazi movement have been busy occasionally attacking and killing Turks (who are mostly Islamic).

In Catholic Poland, where their extremists have also virtually run out of Jews to either kill or hate, they have now invented "designated Jews." That is, those political figures, with whom some members of the populace disagree, have been simply declared to be Jewish by their opponents and then hated on that basis.

In this way, they can be mindlessly and falsely categorized and conveniently and automatically vilified, so that discussions about the real political issues can be easily and entirely ignored.

Various hatreds and intolerance (which has shown up throughout the world, throughout history) have been directed towards specific religious groups (by others) proving the very dangers inherent in the greedy and mindless competition between different religions.

After all, when one values faith over reason, one will believe what they feel like -- and sometimes *especially* when there's nothing rational with which to back it up.

Religion, by its nature is prejudiced and anti-intellectual.

It is religion that spreads hatred either directly or indirectly (sometimes in spite of some very decent and well-meaning individuals who themselves are religious). The fact that these "in groups" each claim to know "the truth" (though none actually do), is enough to re-enforce the justification, in the minds of the individuals, that they are superior in a 'godly' way.

They see themselves as being especially superior to the other heathens who have defiled the one and only *real* god (from among the nearly countless numbers of fake ones) by not falling in line with their religion's dogma.

There are, unfortunately, sometimes rewards for irrational thinking and blind mindless dedication and obedience to whatever one is told. Often honesty, intelligence and truth are totally devalued while being replaced with hypocrisy, lies and fundamental insanity.

It is easy for most people to look at the rituals, for instance, of those who engage in voodoo and to conclude, rightly, that these are mere superstitious practices. Yet how many people are willing to use the same objective reasoning when it comes to looking at "their" own religious views and practices?

After all, what is the difference between voodoo chanting and any other prayers? Or between lighting voodoo candles or lighting any other religious candles?

All religions are virtually identical in that respect.

They appeal to the simplest of minds with the simplest of explanations for some of the world's most complex phenomena. And they all revel in bizarre, pointless ceremonies.

Most people find it easy to follow the masses, blend in mindlessly, and engage in rituals having no genuine function other than ensuring obedient conformity.

It is often a combination of psychological, sociological and political manipulation and control of those who are too worn out mentally even to recognize what is going on to them personally as a result of the actions of this institution.

Why should we respect it?

False claims that poison the minds of the masses should be countered by vigorous intellectual analysis and critique and rewarded with appropriately stark honesty – no matter how annoying the conclusions may seem to some people.

The idea of respecting the individual religious person is perfectly fine. I *do* respect all people, of all faiths – unless and until I'm given a reason not to. But *systems of belief,* the religions themselves, especially those that perpetuate prejudice, injustice, irrationality, ignorance, abuse, lies and anti-intellectualism, deserve contempt, not respect and certainly not support.

But one might ask: what about the individual believer? Do they not have that right, and if it makes them feel better, what's wrong with that?

First of all, whether or not one has a "right" to believe, has nothing to do with whether or not one's beliefs are true. This essay is interested in arguing about the validity, or lack thereof, of religious ideas. The "right to believe" is discussed in the chapter: *Everyone's Entitled To His Own Opinion?*

Secondly, one might counter with the issue of "if it makes them feel better, what's wrong with that?"

The same question can be asked of the drug addict, and for exactly the same reason. It is no coincidence that Karl Marx called religion "the opiate of the masses." In both the cases, the addict only "feels better" because he has become sickened by his addiction in the first place. Otherwise, neither the drug, nor the religion, would have been able to create the illusion of comfort and the false sense of security.

Clearly, there is a matter of individual, subjective valuation involved here. But I don't understand how anyone in good conscience would allow himself to cling onto any belief system either knowing that it is not valid, or not caring whether it is.

Yet the larger issue here is the truth, or lack thereof, behind the odd claims of religion itself – and *not* whether or not people are 'entitled' to hold these beliefs.

Ultimately, I suppose if anyone honestly believes he would feel better spending his entire life pretending to be Napoleon then perhaps that's his right.

But it won't make him Napoleon.

Who Do You Think You Are?

On the surface, this may seem to be a peculiar question to ask. But it's a fundamental question that everyone should ask their self and answer completely; yet very few of us ever *really* do.

Most of us don't fully know who or what we are. This is meant literally and figuratively. We don't seem to know *who* we are when we refer to the apparent 'possibility' of reincarnation, or of an 'afterlife.' And we don't seem to know *what* we are, when we talk of spirits and souls (other than as a metaphor for human personality or conscious awareness).

For example, what exactly is it that would be 'reincarnated'?

Would it be this magical 'soul' that people keep pretending exists? Would it be our undefined 'spirit'?

Or, wouldn't it really just be our conscious awareness?

Of course, when people talk about reincarnation, what they really mean is *the continuation of their specific individual conscious-awareness.* For what if we were 'reincarnated' as a tree, totally unaware of our wooden existence, or of existing in any form whatsoever anyway, then what difference would it make? In what way would there be any meaning to the term "reincarnation"?

Reincarnation, without the continuation of our specific individual conscious awareness, would be totally meaningless for us – and especially for anyone who detests the concept of our existence being absolutely dependent upon our material body. For if we were a material body of any sort, whether a tree, or a rock, or anything *which did not have an awareness of its own existence,* then none of us would know we existed.

In such a case, our existence couldn't possibly matter. We wouldn't know about it, so we wouldn't be able to care about it. At that point, we may as well not exist at all (that is, from our point of view, which, of course, we would be totally missing)!

A tree doesn't even have a central nervous system, much less a functioning brain, and so there is no possibility that it can be consciously aware of its own existence. Likewise, it would make no sense for trees to have evolved which *did* have a conscious awareness. What good would *that* do? What would be the point?

Such a tree would probably be thinking to itself something like, "This is a crashing bore being a tree. My life is tedious and intolerable. I can't do anything but stand here and grow . . . *very, very* slowly. I'm stuck watching birds making their home out of my hair and limbs. 'Cute' little acorn-eating rodents with bushy tails are constantly infesting me. I am frequently humiliated by being soaked in random dog urine."

Once again, if a tree really *did* have conscious awareness, then that would still only prove that what people mean when they used the concept of reincarnation was nothing more than the idea of the continuation of that individual's consciousness into another living being after one's own death, and no "soul" need be involved.

Yet, as we've seen, trees do not have any conscious awareness of their own existence, much less of anything else. They have no sensory apparatus with which to *consciously* detect sights, sounds, heat, pressure, position, tastes, smells, etc. They have no brains with which to interpret such missing data, or with which to have either a thought processes of any kind, or any emotions whatsoever.

Yet they are alive.

Most humans, on the other hand, do seem to have some apparent measure of an awareness of their own existence (at least from time to time). There's good reason for this. People are mobile; trees are not. People can do something about whatever dangers there may be out there in the real world by using their sensory awareness and thought process and then taking whichever actions are perceived to be necessary (thus forcing "free will" decisions).

Which brings us to the standard question: What if one could be reincarnated as another person?

This doesn't even begin to make sense when examined carefully. For *if* reincarnation were a fact, then it would have to be the case that the person's identity (i.e., their thoughts, attitudes and feelings – all that goes into making up an individual's conscious awareness) would be continued into another person's life.

But if it were another person's life, it would have another person's identity – including *their* thoughts, attitudes, feelings and conscious-awareness. Therefore, it wouldn't be yours. If it were yours, it wouldn't be another's.

All of the qualities associated with an individual's specific conscious being are absolutely connected to the specific individual's physical body including their brain. This brain is totally physical and doesn't just inexplicably "vanish" upon death, but slowly rots (or is burned).

There is no mystery about death.

If, when someone died, they disappeared and we could no longer locate them, then there'd be a mystery regarding what happens after we die. But we know what happens to dead people. They rot, get cremated, used for body parts, etc.

And if you think they're not really dead, then why collect on their insurance? Why visit their graves if they're not dead or if you think it's not really them in their graves? [7]We know damned well what death is and what it means – but many people dishonestly refuse to acknowledge this primary fact of life.

Furthermore, every individual's brain is so specific, due to the genetic material, the specific structure and functioning of it (electrochemically), its location in space-time, and the very many specific influences from its environment (including whichever chemicals are ingested by that individual that penetrate the blood brain barrier), that each person's conscious awareness is totally separate from everyone else's. Those are what give us our individual sense of specific identity.

For example, while we may all be looking at things which are objectively

---

7. Of course, why do that anyway?

real, and whose existence is completely independent of our own, the fact that, as of yet, none of us can share our individual conscious awareness with any others of us is what gives us the strong illusion that we have some magical component (a 'soul' not made of anything physical and not specifically located), that somehow separates us from everything else in the universe. This leads some people to believe that we can survive our death and be installed in someone else's brain (who, presumably, doesn't already have one of these 'souls').

Oh, we can share objects whose independent existence we can each perceive separately, but we can't share our specific conscious awareness. I can give you an apple which I see, and you can see the same apple at the same time, but your awareness of the fact of your seeing that apple is a separate thing from my awareness of the fact of my seeing that same apple.

Even though we both see the same apple, and are each aware of seeing the same apple, neither one of us is aware of the other's individually specific awareness.

What, then, is it *exactly* that makes you "you" and me "me"?

To understand this is to define it.

We can define ourselves any way we want, but for all reasonable and practical purposes, we might wish to define ourselves in a manner consistent with how we generally tend to (unconsciously perhaps, and in unspecific ways) define other people. That is, we are the sum total of everything that we've been from the moment of first conscious awareness to the moment of our final conscious awareness.

In other words, while we would certainly answer "Shakespeare" when asked "Who is buried in Shakespeare's grave?" that doesn't mean that we think of Shakespeare as being mostly dead (even though he's been dead for a far longer period of time than he was ever alive). No. Indeed we think of everyone as being alive – even when we know they are no longer living.

That is why people say things like "My (great) grandmother would have been 100 years old today," when, in reality, she *is* 100 years old (if counting from her date of birth).

The only difference is that she's also dead.

But if this were the hundredth anniversary of her birth, she'd indeed be one hundred years old regardless of her current condition; which brings us back to our initial question of what is it that exactly, precisely defines an individual.

Once more, the above example is anecdotal evidence that we do tend to think of people as being defined by the scope of time bounded by their first conscious awareness (usually birth, sometimes before) to their final conscious awareness (usually death, sometimes before).

The answer to the question, "whose bones are these?" may well be "they are Shakespeare's." Thus acknowledging that, indeed, they are Shakespeare's bones. Yet we also stop thinking of Shakespeare *as being Shakespeare* the moment of his death. Naturally, this applies to everyone else as well.

So what does this mean?

It means that we are defined within, and by, our own lifetime, even though we know that *everything that makes us up has existed before our birth and continues existing, in one form or another, forever after our death.* As noted before, we don't mysteriously vanish upon our death, and everyone recognizes who is buried in whose grave, thus acknowledging that the person is dead.

We can't very well have dead people without also having them be the specific individuals who are dead, can we?

So we recognize that our confinement of the individual to coincide with his/her conscious awareness is arbitrary. Nevertheless, it's a fair, if convenient, way to quantify a person's existence.

Now we must ask, "What is it exactly that makes us up?" In other words, aside from the limit of defining us as being, "from the 'first' conscious awareness to the 'final' conscious awareness of the specific person," what is that continuous entity of conscious awareness made of?

We are made of oxygen, hydrogen, carbon, nitrogen, etc.

In fact, all of what makes us up is both non-living electro chemical energy-matter, and also has always existed and will always exist in one form or another.

Part of your hand may have once been part of one or more dinosaur. Certainly, if you have ever eaten bacon and eggs, then at least at some point during your lifetime, parts of you had once been parts of a pig and a chicken. If you stop breathing you will die in a very short period of time due to lack of oxygen. Therefore, it is perfectly reasonable to understand that bits and pieces of the air you have yet to breathe will become bits and pieces of the future you.

Yet we tend to ignore the obvious when defining ourselves.

We want, so desperately, to be so very unique and special that we just ignore reality when it's convenient to do so (which, for most people, may well be most of the time). But if you really want to truly understand something, you have to face it directly and explore it in all of its logical detail to actually know what it is (to the degree that it can be known at all).

The Problem With Infinity and Other Flaws in Mathematics

The very foundation of the legitimacy and validity of mathematics is based on the claim that it is internally logically consistent, has clear meaning, and is "true."

While these three assumptions are not entirely wrong, they are also not entirely right. As a result, the consequence of following (and nearly worshipping) mathematics, and endowing it with more logic than it really has, has caused major problems for the scientific community which border on near-religious hysteria when dealing with the philosophical consequences of issues on a sub-atomic scale.

Zeno's never-ending paradox. . .

According to Zeno's paradox, and in accordance with the principles of mathematics, if a tortoise proceeds from a starting point (point A), to the finish line (point B) [or, if a coin drops from a distance (starting at point A) to the table or floor (ending at point B)], there are how many points between A and B? The answer, according to mathematics, is an "infinite amount" (sic) of them.

But before the tortoise reaches the finish line, or the coin hits the floor (or table), it would have to go halfway between where it started and where it would end. But before it could get to its end point B, it would have to go halfway between the halfway point and the end point, etc. And since there are an "infinite amount" (sic) of points, it should always be traveling from halfway to halfway, and should never reach the end.

Yet it does.

And that's Zeno's paradox.

The resolution? (Since there's no such thing as a paradox that cannot be solved rationally...) Well, we know these things about this:

1. It's true that mathematics claims there are an infinite amount of points that anything would have to traverse between point A (the start) and point B (the finish) since, by definition, that would be the course of a line segment which has an infinite amount of points.

2. But it does land at the end point.

3. Therefore, mathematics is not an accurate description of reality and in reality there is nothing infinite whatsoever (or the coin would be forever falling and the tortoise would never reach the finish point).

This is not a matter of semantics, but a matter of fact.

Once a concept has been proven false, it can no longer be regarded as being true. Though this is the simplest way of destroying infinity (in addition to pointing out the inherent flaw in the mutually-contradictory term "infinite amount"), it remains a valid one.

We know what infinity would be if it existed. But we also know that if it did exist, things could never move, much less get to their destination because it would always be taking forever to get from one (imaginary) point to the other since there would be an "infinite amount" of points between them and they wouldn't be able to cover all of these points or they (the points) would not be infinite (ongoing and not ever coming to an end). Either it comes to an end (which *is* the case) and therefore is not infinite, or it never ends/lands/reaches its goal and therefore is infinite (which *isn't* the case).

But things of that nature do reach their destinations, so while the concept of infinity may be useful, it is not accurate, as it does not truly describe the real world (universe).

Line Segments

In mathematics, a line and/or a line segment is defined as having one dimension (let's call it "length"). A plane is said to have two dimensions (length and width). A cube is said to have three dimensions (length, width and depth). Yet for something to exist in reality, it must have all (at least four) dimensions – the three spatial ones, plus some duration in time. (Super string theory would not interfere in the logic I present here, so if there were other dimensions, they would all exist equally at once and for everything).

Points are defined as having "location" in either time or space, but having zero dimensions. That is, while lines and line segments are defined to have one dimension (for example, "length"), points are defined to have none.

No dimension.

Zero.

Let's take a line segment. Rather, let's imagine one. (There are no such things in real life, of course, but we tend to think we know what a "line" looks like,[1] so we can imagine a line segment, which, unlike a line itself, would have both a beginning and an end.)

A line segment is, by definition, supposedly made of an "infinite amount"[2] of "points." "Points" have no dimension—again, by definition. Therefore, it follows logically that no matter how many points we put together (which is the same as multiplying them, obviously), we will only have a single point of

---

1. Of course we can neither imagine a "real" line, nor a line segment. In the first case, we cannot imagine a line because we would always only be visualizing part of it – since it has neither a "beginning" nor an "end." Another reason we cannot really imagine a line or a line segment is because neither have any more than one dimension (say, for example, "length"). Therefore, any imaginary thickness we would "see" in our minds would not ever be part of any "real" line or line segment.

2. The term "infinite amount" is used, because there is no other kind of term to put in its place, though we recognize that amounts require definite numbers in order to have any real meaning, and that the word infinite is not itself an amount (or quantity of any kind in any way), but merely a word which means "never ending." Therefore, the phrase "infinite amount," while inherently contradictory, is apparently 'understood' by everyone, so is used here in common parlance.

no dimensions whatsoever. After all, zero anything (in this case, dimensions), times anything at all will yield only zero. That is what mathematics tells us. That is one internal logical conflict right there.

Since there can be no "infinite amount" of points yielding a line segment (much less an entire line), it is clear that a line segment cannot be said to be made entirely of an infinite amount of points. So into what kind of smallest quantities *can* it be broken down? It can only contain smaller line segments, but not an "infinite amount."

Therefore, a *finite number of smallest possible line segments* is the only ultimately irreducible division of any given imaginary line segment itself.

And that "smallest possible" line segment has to be greater than zero in length (or it would have no line-like qualities; i.e., it wouldn't be a line segment itself but a point of no dimension and therefore you could keep adding them up forever and you wouldn't get anywhere further in the dimension so no line segment could ever be made of these things, no matter how many there were).

As for the line, or line segment, itself, no such thing as an "infinite amount" of anything can exist since it depends upon infinity for its definition and, as we've seen, the concept of infinity is itself irrational and not logically possible to exist in the universe outside of being a concept in math.

Odds and Ends

Everyone's Entitled to His Own Opinion – ?

Once more we encounter and decipher an old cliché frequently used in place of, for the purpose of avoiding, a rational debate (as is true of the use of the phrase "that's just common sense").

Put simply: exactly what "entitles" us to our own opinion (whatever that is)? Is it simply the fact that we have one? For if having something "entitles" us to it, then there'd be no such thing as thievery.

First, what do we mean by "entitlement"?

Evidently this is the idea that we have some agreed upon societal right to "own" (possess) "our own" (our unique, though, of course, it isn't) opinion. But do we? And if so, why? How would that assumption be justified?

I know we may have a right to believe, *a priori,* that we're "entitled" to our own opinion – and maybe we are (or should be). But why do we take this for granted? What makes it a given?

If you were of the opinion that people wearing funny hats should be killed on sight and were willing to do it yourself because of that very strong opinion, are you be *entitled* to such a view? *Should* you be?

Now, I know people will say, "yes," as long as one doesn't act on it. But in my example, one *does* act on it, and wouldn't have done so without having had such a fanatical opinion to begin with. Remember, thoughts that are acted upon are causal agents (which is the same as forces) in the ever-occurring chain of events of universal causality.

Of course, some of you will say that I've already argued that we live in a super-determined universe where everything not only has to happen exactly as it does, but that it already has and is just waiting for us to experience our part in all of this – therefore, how dare I assign blame for anything anyone does when it was always inevitable? . . . Where does responsibility lie in a totally determined world?

My answer is that we are *de facto* responsible, in any case, and will take the consequences of our behavior whether we are to "blame" or not. Furthermore, we have conscious awareness and can think before acting (the thinking being part of the causal chain of events eventually forcing us to behave as we do) so we should consider the consequences of not only our actions, but also of the thoughts behind those actions – even before we are ever aware of the potential actions themselves.

Regarding whether or not we're "entitled" to "our own" opinions, this is a purely subjective issue. If one were "entitled" to anything, it would be by the agreement of the society that entitles us.

On the other hand, if we think of "entitlement" in a personally subjective way (such as meaning "deserving") then it's up to each of us as to what we believe we are "entitled" to.

Therefore, you'd get to decide this one for yourself (of course, others might disagree).

After all, you're entitled to your own opinion.

The Silly Philosophical Stuff –
From the sublime to the ridiculous and vice versa.

Since I started this book by discussing the simple concepts of nothing and nowhere, I thought I'd end it with some more simple concepts.

If a tree falls in the forest, do you really give a damn?

Here comes the fun stuff.

Philosophy has always included, along with the profound and deep, the absolutely trivial and pseudo-intellectual crap that passes itself off as being profound, but in reality is curiously devoid of substance.

Here then are some of the life-long questions and (apparent) paradoxes, along with the logical explanations that lead to their answers.

If a tree falls in the forest, and no one is there to hear it, does it make a sound?

The answer depends entirely upon how one defines the word "sound." Usually, this word is defined as "the physical phenomenon which occurs when air molecules are vibrated;" these vibrations reach the inner ear and are then transmitted to the brain, through the central nervous system, where they are interpreted (or "heard"). So, by that definition, if a tree falls in the forest and nobody is around to 'hear' it, it certainly *does* make a sound.

If, on the other hand, the word "sound" is defined to mean simply "that which is heard," then, by that definition, if no one hears something, that something cannot be considered to be a sound. This is strictly a matter of definition, and nothing else. (Of course, the "one" that hears the sound need not be human.)

There really is something else to this question in the minds of some people. That is, they may be willing to define the word sound as being the vibration of air molecules, but think that perhaps if we cannot be there to observe an event, that this event hasn't really happened.[1]

Their internal logic contradicts itself. For an event such as "a tree falling in the forest" is already a given in this example. The very people who postulate this real tree falling in a real forest, then claim that if we aren't there, it didn't happen. Well, you can't claim something happened and then claim that it didn't in the same breath and pretend that anyone with a functioning brain is going to take you seriously

So now we know, depending upon whether or not a sound is defined as something that needs to be heard, in order to be considered a sound, will determine whether or not that proverbial tree falling in the forest produces a sound.

---

1. On the face of it, this idea is strictly irrational. However, in the world of physics, on the sub-atomic level, the way we observe certain kinds of events actually does influence the events themselves. These phenomena are what tend to support such odd ideas in the minds of some people. Yet "scientists" who aren't being very scientific in their analysis of these events misinterpret even this. For it's a lot easier to inadvertently disturb subatomic particles by, for example, bombarding them with photons, than it is to move a parked car by exposing it to the Sun.

Which came first, the chicken or the egg?

This is an easy one. The answer is, naturally enough, "the egg."

Why?

To begin with, whatever one wishes to label as being "the first chicken," is going to be an arbitrary decision. (And at some point, someone who asks "which came first" will have to claim that there was a "first chicken.")

Chickens, like everything else, evolved. [In truth, there really is no first chicken because the line between one species and another is not a sudden sharp one. Evolution in species occurs over relatively long periods of time (dependent upon the life span of the animal, naturally) and happens gradually.]

Yet if we wish to impose a "first" chicken, then it will have to be an arbitrary label – but one which, nevertheless, will have to conform to all essential chicken-like characteristics, one of which is that chickens, not being mammals, are birds and birds are egg-laying creatures. Therefore, the "first" chicken would have come from an egg-laying creature (a pre-chicken, if you will), and that means the egg (that the first chicken hatched from) came first.

What happens when an immovable object meets an irresistible force?

Again, this apparent paradox is based on a false premise; that is, that there can be such things as "immovable objects" and "irresistible forces" and that they can peacefully exist together in the same universe. This is a logical contradiction. The very definition of an "immovable object" is such that this object cannot be moved by any force or means whatsoever. The definition of an "irresistible force," is a force that cannot be resisted by anything (that is, all objects would have to be moved by it). They are mutually exclusive.

Naturally, neither thing could *really* exist in the universe, since both "objects" and "forces" are physical entities obeying physical laws and would eventually meet up in the *Big Bang / Big Crunch* near-singularity (as mentioned in a previous chapter). In addition, whether or not an object is "at rest" or is "moving" cannot be determined in any absolute manner. All motion of all kinds is relative. That is, things can only be said to be moving by comparing them to something else (with respect to which they are either moving towards, away from, or are rotating about). Of course, the very fact that these are two mutually exclusive concepts means they can't both exist together anyway.

But . . . what if they could?

Is there a way to salvage the idea of an irresistible force being able to exist along with an immovable object, let's say as a "thought experiment"? That is, can we imagine any scenario in which this might not be a logical conflict? The answer, surprisingly enough, is yes, sort of (but only if we cheat a little).

Let's imagine the immovable object to be possible (which it isn't – but we can pretend for a moment). Let's make this object a wall somewhere "anchored" permanently in the universe. Let's say that it is this wall that cannot be moved by any force whatsoever.

Now, let's deal with the imaginary irresistible force. Let's pretend this force propels a sphere made of the same amazingly tough material out of which the wall is made.

We can further visualize this "irresistible force" propelling the sphere through space. Everything that gets in the way of this projectile is pushed aside like a bullet through butter. Nothing resists the irresistible force. So what would happen were it to meet up with the immovable object?

The answer: it would bounce off – without losing any energy (so it would still not have been resisted so much as it would have been re-directed, depending upon the angle at which it hit the wall, back where it came from). The wall doesn't move, keeping it immovable, and the force isn't resisted so much as it is re-directed like a laser light hitting a spot on a mirror.

"Ah," you may say, "but that would be cheating."

To which I reply: "The only cheating involved is pretending that such things could be real, but if they were, the rest is not cheating. Here's why: *the sphere*, which the irresistible force propels, does not cause the wall to move, of course. So you might think that it hasn't satisfied the requirements of the original condition of the thought experiment. But, it isn't *the sphere* to which we are referring, it is the *force propelling the sphere* – and that isn't ever being resisted by the sphere at all."

So you see, it can be accomplished -- even if only in a thought experiment.

What is the sound of one hand clapping?

There are several possible responses to this one – all pretty much equally silly and yet all appropriate. The first response, "it is half the sound of two hands clapping." The next response was demonstrated on an episode of the cartoon television show *The Simpsons* (though it was not original with them) where Bart Simpson continually applauds with one hand by rapidly repeating the motion of bending and extending his fingers to hit his palm.

The point of a Zen question, of course, is to ask the question itself, and not necessarily to find an answer (which may or may not exist – depending upon whether or not the question has any real meaning). At least, that is what I think– however, I don't care for rhetorical questions. I don't allow them to go unanswered by me and I never ask them of anyone else.

Why would I?

A Brief Comment on Uniting Gravitation
With the Electro-Weak Forces

There is a current problem in modern physics. It is the seemingly contradictory and mutually exclusive natures of two theories: Einstein's General Relativity, and Quantum Mechanics.

Many people are busy searching for some means by which to unify these.

The Unified Field Theory (or, Theory of Everything), as it is called, aims to resolve some seemingly inherent conflicts between General Relativity and Quantum Mechanics. That is, how do we explain every physical phenomena under one theory, when the two best ones we currently have, seem to have irreconcilable differences?

General Relativity *explains* things on the macro (large) scale, while Quantum Mechanics *predicts outcomes of events* on the sub-atomic scale.

So the problem of unifying the two is odd because they do different things in different realms. One explains philosophically (while proving mathematically) how the universe works in the large scale, and it is mechanistic and causal. The other doesn't explain anything at all. It just predicts large numbers of events in the sub atomic world with remarkable accuracy, and (this is important), with no valid accurate philosophical explanation and interpretation of what exactly is happening and why.

In other words, the very justification for taking Quantum Mechanics seriously is our understanding that it comes closer to predicting, with near total accuracy, what will happen in the sub atomic world. And it does so better than any other mathematical model. But the very same factor that gives QM its legitimacy is exactly what the bad philosophical interpretations of it take away.

That is, the assumption that there is such a thing as causality to begin with and those events will occur in specific and definite and ultimately, in theory, predictable ways if we had all the variables.

Even if we don't have access to (and never can have access to) all of the variables [remember the example of the all-knowing computer in the completely deterministic universe?], the idea is that they do exist or else why give credit to a system (QM) for coming closer to predicting with total accuracy -- if the concept of total accuracy is false? That would be self-defeating as well as being logically contradictory.

Yet, the larger world is made directly of the smaller one. So we know the physical laws have to be united.

We need to examine the bad interpretation of the *Heisenberg Uncertainty Principle* known as the Copenhagen Interpretation.

It is true that one cannot set up an experiment that will *both* locate a sub-atomic particle, such as an electron, and at the same time tell us its velocity. That is, either we can find out how fast it's moving (during a given range of time), or else we can discover *where* it is (at any given "moment") – but not both at once.

This is a fact.

What isn't a fact, however, is the faulty reasoning associated with the Copenhagen Interpretation of the Heisenberg Uncertainty Principle. Basically, the Uncertainty Principle itself says that because we can't know *both* the location *and* momentum of a sub-atomic particle at the same time, that one of those facts will forever be uncertain to us. And once more, this is true.

But the Copenhagen Interpretation of that fact arbitrarily places the uncertainty on the event itself, rather than on the observer where it belongs.

[And, by this point in the book, it should be sharply clear that events cannot be uncertain in a completely deterministic universe – where not only is the future totally determined, but it's already in existence. Nor would it make sense to believe events themselves can be uncertain. That would be like claiming it's "cold outside" – how could it be? You could be cold outside – the outside itself, however, cannot be. As we've already discussed, all events either *do* occur or they *do not* occur. So the events themselves can, in no way, happen in any uncertain manner.]

Chaos, Order And The Meaning Of Life

A little note on chaos (randomness) and order:

The difference between what we call "chaos" and what we call "order" is simply this: Order is that which we find useful. Chaos is that which we find messy, destructive, and, aesthetically out of place. It's nothing more than that.

The "chaotic" results of an explosion, or of the path and effects of a volcanic eruption, or of the unpleasant product of a night of drunken debauchery gone wrong, ending with a vast stream of projectile vomiting, are truly neither "random" nor "chaotic." They are no more so than a totally predictable, but convenient, occurrence such as the next full moon, or deciphering how much thrust is needed to send a specific rocket off into outer space.

In all cases we are talking about events that are equally subject to the laws of the universe (or laws of Nature, or laws of physics – however you'd care to label them). But the same kind of unexamined human prejudices that cause us to pretend that if we're cold outside then "it's" cold outside – are the same kinds of prejudices that will lead us to arrogantly and falsely conclude that because we don't have all of the facts, then not all the facts exist and therefore things can be random and chaotic and are not occurring in specific, exact and predetermined ways.

The truth is, of course, that even though what we call "chaos" or "randomness" is extremely complex (and generally useless to us), and the specific pieces of information (variables) necessary to properly predict them may be out of our grasp, all events follow physical laws equally for the frames of reference in which they occur and all events will happen exactly as they do.

[Of course by this point in the book, all of this should be self-evident.]

Now, let's proceed to the meaning of life...

The meaning of life ...

The word "meaning " is, by its nature, subjective.

In order for anything to have meaning, it must mean something to someone (or to some conscious entity). Just as nothing can have "intrinsic value," neither can it have "intrinsic meaning."

Meaning, or purpose, absolutely requires someone for whom the thing under discussion has meaning. For us, the meaning of our lives depends upon who is asked the question. Certainly, the meaning of your life to a total stranger is nothing at all. They don't even know you exist, so naturally your life will have no meaning whatsoever to them. To your friends, family and (presumably) to yourself, however, your own life will have a great deal of meaning.

But what does it mean?

Again, it depends on whom you ask. For you, your own life should have the greatest possible meaning. And the good news is that what it means is entirely up to you.

But what meaning is there regarding life in general? What is the meaning of that?

Again, the answer depends on the person (or sentient being of any kind, living or otherwise), who is asked that question. And far from being unimportant, it is the most important question one can ask of oneself, and the answer is up to each of us to decide.

May we each decide wisely.

Postlude

There are many more ideas in the realm of philosophy that I did not include in this book (such as delving more deeply into the spheres of linguistics, ethics, and epistemology).

I plan on writing more books of a philosophical nature in the future (as this is the only one I've written in "the past" – so far). But I think this has been a suitable start. I would like to add a few words about comfort and intellectual exploration.

Comfort's best when grounded in reality and when that reality is told by and to people who have a strong regard for the truth, reason, and for the assumed dignity of humankind.

Don't be afraid of performing the occasional "mental house-cleaning," by picking up the contents of your mind, holding them up to the light, examining them carefully, and throwing away the crap.

I hope you have enjoyed this book and have learned from it. I also hope it has stimulated your brain to function rationally within a high moral/ethical framework that causes you to dedicate yourself to the continuing improvement of your own life and of society at large.

Regards,

Michael Perilstein

Notes

Notes

Notes

Notes

www.ingramcontent.com/pod-product-compliance
Lightning Source LLC
Chambersburg PA
CBHW080511110426
42742CB00017B/3078